Generation and Applications of Extra-Terrestrial Environments on Earth

RIVER PUBLISHERS SERIES IN STANDARDISATION

Volume 6

Series Editor

Anand R. Prasad
NEC, Japan

The "River Publishers Series in Standardisation" is a comprehensive series addressing the pre-development related standards issues and standardized technologies already deployed. The focus of this series is also to examine the application domains of standardised technologies. This series will present works of fora and standardization bodies like IETF, 3GPP, IEEE, ARIB, TTA, CCSA, WiMAX, Bluetooth, ZigBee etc.

Other than standards, this book series also presents technologies and concepts that have prevailed as de-facto.

Scope of this series also addresses prevailing applications which lead to regulatory and policy issues. This may also lead towards harmonization and standardization of activities across industries.

For a list of other books in this series, visit www.riverpublishers.com
http://riverpublishers.com/bookspubseries.php?sn=Standardisation

Generation and Applications of Extra-Terrestrial Environments on Earth

Editors

Daniel A. Beysens

CEA-Grenoble and ESPCI-Paris-Tech,
Paris, France

Jack J. W. A. van Loon

VU-University, Amsterdam,
The Netherlands

Published, sold and distributed by:
River Publishers
Niels Jernes Vej 10
9220 Aalborg Ø
Denmark

ISBN: 978-87-93237-53-7 (Hardback)
978-87-93237-54-4 (Ebook)

Contents

4. Other Environment Parameters

5. Current Research in Physical Sciences

Preface

This book has been prepared under the auspice of the European Low Gravity Research Association (ELGRA). As a scientific organization the main task of ELGRA is to foster the scientific community in Europe and beyond in conducting gravity and space related research.

This publication is dedicated to the science community, and especially to the next generation of scientists and engineers interested in space researches. ELGRA provides here a comprehensive description of space conditions and means that have been developed on Earth to perform space environmental and (micro-)gravity related research.

We want to thank all our colleagues who contributed to the interesting and hopefully inspiring content of this book. It is the first in its kind to addressing a comprehensive overview of ground-based technologies and sciences related to (micro-)gravity, radiation and space environment simulation research.

Daniel A. Beysens and Jack J. W. A. van Loon

List of Contributors

Dr. Carmen Arena *University of Naples Federico II, Dept. Biology, Naples, Italy*
Email: carena@unina.it

Prof. dr. Giovanna Aronne *University of Naples Federico II, Dept. Agricultural and Food Sciences, Portici (Naples), Italy*
Email: aronne@unina.it

Miss Marie-Pierre Bareille PharmD *MEDES, Toulouse, France*
Email: marie-pierre.bareille@medes.fr

Prof. dr. Jerry Joseph Batzel, PhD *Gravitational Physiology and Medicine Research Unit, Institute of physiology, Medical University of Graz, Austria*
Email: Jerry.batzel@medunigraz.at

Prof. dr. Daniel A. Beysens *CEA-Grenoble and ESPCI-Paris-Tech, Paris, France*
Email: daniel.beysens@espci.fr

Christian Chauveau *CNRS – INSIS – ICARE, Orléans, France*
Email: christian.chauveau@cnrs-orleans.fr

Prof. dr. Marco Durante *GSI Helmholtzzentrum für Schwerionenforschung and Technische Universität Darmstadt, Darmstadt, Germany*
Email: m.durante@gsi.de

Prof. dr. Christoph Egbers *Brandenburg University of Technology Cottbus, Germany*
Email: egbers@tu-cottbus.de

Prof. dr. Pascale Ehrenfreund *Leiden Observatory Leiden, The Netherlands*
Email: p.ehrenfreund@chem.leidenuniv.nl

Prof. dr. Hans-Jörg Fecht *Ulm University, Ulm, Germany*
Email: hans.fecht@uni-ulm.de

Dr. Birgit Futterer *Brandenburg University of Technology Cottbus, Germany/Otto von Guericke Universität Magdeburg, Germany*
Email: birgit.futterer@web.de

Dr. Vasily Gnyubkin, *INSERM U1059, LBTO, Faculty of Medicine, University of Lyon, Saint-Etienne, France*
Email: vasily.gnyubkin@univ-st-etienne.fr

Prof. dr. Nandu Goswami M.D., PhD *Gravitational Physiology and Medicine Research Unit, Institute of physiology, Medical University of Graz, Austria*
Email: Nandu.goswami@medunigraz.at

Prof. dr. Karl H. Hasenstein *Biology Dept., University of Louisiana at Lafayette, Lafayette, LA – USA*
Email: hasenstein@louisiana.edu

Dr. Raúl Herranz *Centro de Investigaciones Biológicas (CSIC) Madrid, Spain*
Email: rherranz@cib.csic.es

Dr. Richard J. A. Hill *School of Physics and Astronomy, University of Nottingham, UK*
Email: richard.hill@nottingham.ac.uk

Dr. ir. Inge Loes ten Kate *Department of Earth Sciences, Utrecht University, The Netherlands*
Email: i.l.tenkate@uu.nl

Prof. dr. Yasuhiro Kumei *Tokyo Medical and Dental University, Japan*
Email: yasuhirokumei@gmail.com

Prof. dr. Claus Lämmerzahl *Center of Applied Space Technology and Microgravity (ZARM), University of Bremen, Bremen, Germany*
Email: claus.laemmerzahl@zarm.uni-bremen.de

Dr. Natalie Leys *Unit for Microbiology, Belgian Nuclear Research Center (SCK•CEN), Mol, Belgium*
Email: Natalie.Leys@sckcen.be

Dr. Ing. Jack J. W. A. van Loon *DESC (Dutch Experiment Support Center): Dept. Oral and Maxillofacial Surgery/Oral Pathology, VU University Medical Center & Dept. Oral Cell Biology, Academic Centre for Dentistry Amsterdam (ACTA) Amsterdam, The Netherlands*
Email: j.vanloon@vumc.nl

Dr. Clement Lorin *Mechanical Engineering Department, University of Houston, USA*
Email: clement.lorin@gmail.com

Dr. Alain Mailfert *Laboratoire Géoressources, CNRS-Université de Lorraine, France*
Email: alain.mailfert@univ-lorraine.fr

Dr. Alain Maillet *MEDES, Toulouse, France*
Email: Alain.Maillet@cnes.fr

Dr. Felice Mastroleo *Unit for Microbiology, Belgian Nuclear Research Center (SCK•CEN), Mol, Belgium*
Email: fmastrol@sckcen.be

Dr. F. Javier Medina *Centro de Investigaciones Biológicas (CSIC) Madrid, Spain*
Email: fjmedina@cib.csic.es

Dr. Veronica De Micco *University of Naples Federico II, Dept. Agricultural and Food Sciences Portici (Naples), Italy*
Email: demicco@unina.it

Prof. dr. Greg Morfill *Max Planck Institute for Extraterrestrial Physics, Garching, Gremany*
Email: gem@mpe.mpg.de

Dr. Raheleh Motamedi *Department of Earth and Life Sciences, VU University, Amsterdam, The Netherlands*
Email: raheleh.motamedi@gmail.com

Dr. Innocent Mutabazi *LOMC, UMR 6294, CNRS-Université du Havre, France*
Email: innocent.mutabazi@univ-lehavre.fr

Dr. Ir. Vladimir Pletser *Astronauts and ISS Utilisation Dept, Directorate of Human Space Flight and Operations, European Space and Technology Reseach Center (ESTEC), European Space Agency (ESA), Noordwijk, The Netherlands*
Email: Vladimir.Pletser@esa.int

Dr. Louisa. J. Preston *Department of Physical Sciences, The Open University, Milton Keynes, United Kingdom*
Email: louisajanepreston@gmail.com

Sébastien Rouquette *CNES—CADMOS, Toulouse, France*
Email: sebastien.rouquette@cnes.fr

Prof. dr. Valentina Shevtsova *Université Libre de Bruxelles, Belgium*
Email: vshev@ulb.ac.be

Prof. dr. Theodore Steinberg *School of Chemistry, Physics and Mechanical Engineering, Science and Engineering Faculty, Queensland University of Technology, Brisbane, Australia*
Email: t.steinberg@qut.edu.au

Dr. Carole Tafforin M.D. *Ethospace, Research and Study Group in Human and Space Ethology, Toulouse, France*
Email: ethospace@orange.fr

Dr. Cora S. Thiel *Institute of Anatomy, Faculty of Medicine, University of Zurich, Zurich, Switzerland*
Email: cora.thiel@anatom.uzh.ch

Prof. Dr. Oliver Ullrich *Institute of Anatomy, Faculty of Medicine, University of Zurich, Zurich, Switzerland*
Email: oliver.ullrich@uzh.ch

Prof. dr. Giovanna Valenti, PhD *Biotechnology and Biopharmaceutics, University of Bari, Italy*
Email: giovanna.valenti@uniba.it

Prof. dr. Laurence Vico M.D. *INSERM U1059, LBTO, Faculty of Medicine, University of Lyon, Saint-Etienne, France*
Email: vico@univ-st-etienne.fr

Dr. Harunori Yoshikawa *LOMC, UMR 6294, CNRS-Université du Havre, France*
Email: harunori.yoshikawa@unice.fr

List of Figures

List of Tables

List of Abbreviations

3D	three-dimensional
AMASE	Arctic Mars Analog Svalbard Expedition
ARC	Ames Research Centre
ASC	Atmospheric Sample Chamber
ASUVG	Arizona State University Vortex Generator
AWTS	Aarhus Wind Tunnel Simulator
BDC	Baseline Data Collection
BEC	Bose–Einstein condensation
BISAL	Boulby International Subsurface Astrobiology Laboratory
BLSS	Bioregenerative Life Support Systems
BMI	Body Mass Index
BR	Bed rest
CAB	Centro de Astrobiología
CAFE	Concepts for Activities in the Field for Exploration
CAPSULS	Canadian Astronaut Program Space Unit Life Simulation
CCD	Charge-Coupled Device
CHF	critical heat flux
CNES	Centre National d'Etudes Spatiales (French Space Agency)
Desert RATS	Desert Research and Technology Studies
DLR	DeutschesZentrum fur Luft-und Raumfahrt
DNA	Deoxyribonucleic Acid
ESA	European Space Agency
EVA	Extra Vehicular Activity
EXEMSI	EXperimental campaign for European Manned Space Infrastructure
FMARS	Flashline Mars Arctic Research Station
g	Earth acceleration (9.81 m/s^2)
HDBR	Head-down bed rest
HDT	Head-down tilt
HERA	Human Exploration Research Analog
HI-SEAS	Hawaï Space Exploration Analog and Simulation
HUBES	HUman Behavior in Extended Space flight
HZE	High-energy and charge

IAA	International Academy of Astronauts
IMBP	Institute for BioMedical Problems
IR	infrared
ISEMSI	Isolation Study for European Manned Space Infrastructure
ISS	International Space Station
IVA	Intra Vehicular Activity
JAXA	Japan Aerospace Exploration Agency
LET	High-linear energy transfer
LSMMG	low-shear modeled microgravity
MARSWIT	Mars Surface Wind Tunnel
MDRS	Mars Desert Research Station
MESCH	Mars Environmental Simulation Chamber
MG	Magnetogravitational
MGE	Magnetogravitationalequipotentials
MIM	Microgravity Isolation Mount
MSF	Mars Simulation Facility
NASA	National Aeronautics and Space Administration
NEEMO	NASA Extreme Environment Mission Operations
PAL	Planetary Aeolian Laboratory
PALLAS	Planetary Analogues Laboratory for Light, Atmosphere and Surface Simulations
PS	Photosystem
QS	quorum sensing
R	Recovery
RGA	residual gas analyzer
ROS	Reactive oxygen species
RPM	Random positioning machine
RWV	rotating wall vessel
s	second
SAM	Sample Analysis at Mars instrument suite
SFINCSS	Simulation of Flight International Crew on Space Station
TQCM	thermoelectric quartz crystal microbalance
USA	United States of America
USSR	Union of Soviet Socialist Republics
UV	ultraviolet
μg	micro-gravity

Introduction

Daniel A. Beysens[1] and Jack J. W. A. van Loon[2]

[1]CEA-Grenoble and ESPCI-Paris-Tech, France
[2]VUmc, VU-University, Amsterdam, The Netherlands

The middle of the twentieth century has been the time where mankind got access to space. It has been the beginning of an exciting adventure for cosmonauts/astronauts, engineers, and scientists, where the very specific environment of space led to novel and questioning situations. In the domain of life science, the situation of near weightlessness has strong incidence on, for example, the vestibular system, blood circulation, or bone and muscle degradation. In addition, long stays in space in relatively small confined spacecraft lead for humans to specific psychological conditions that might be compromising for long-duration missions such as to Mars. For physicists, the cancellation of buoyancy in fluids emphasizes the other non-gravity-related forces such as capillary and diffusion effects. Space is also an environment which is characterized by, especially beyond the Van Allen belt, strong radiation and high vacuum.

A large number of studies concerning the comportment of humans, animals, plants, or cells have been conducted in space, as the behavior of matter in the process of solidification, fluid behavior, combustion, etc. However, the access to space is a costly and long process, especially if human presence is required, because either human is the object of study or he is running the experiment. This is why a number of substitutes to space have been elaborated on Earth to prepare or replace the envisaged in-flight experiments.

The basic function of this book is to review and discuss advantages and inconveniences of these ground-based technologies with respect to a real space environment. For this purpose, the first section is dedicated to the description of the space environment: low gravity and weightlessness, atmosphere (vacuum, temperature), radiation and space weather and interstellar chemistry with some considerations on the surface properties of asteroids, moons, and planets. Then, the means to alter weight on Earth are discussed: free fall (drop towers), parabolic flights with planes and sounding rockets, compensation by volume forces such as magnetic field gradients and electric fields, density-matched liquid mixtures (Plateau method) and the use of centrifuges to increase the range of gravity effects.

Zero gravity can also be mimicked by compensating its effects. Regarding the blood circulation and musculoskeleton unloading, animals suspended by their tails or humans in a bed rest, head down tilt are typical means that can be used. Clinostats and random positioning machines are mechanical devices where the orientation of a small sample (e.g., cells, plants, or small animals) is randomly changed with respect to the horizontal, at a timescale smaller than the expected timescale of evolution of the sample. In the same kind of phenomena, vibrating fluids with time period much smaller than the timescale of fluid evolution lead to mean flows that can counterbalance buoyancy-driven flows.

Facilities have been constructed to recreate the conditions of irradiation, or the atmosphere of the space environment in so-called Earth analogues. The psychological problems that individual human beings of a group of persons can undergo in the long periods of confinement in space can be reproduced in near-inaccessible areas such as submarines or on Antarctic stations.

All these means suffer from one or several drawbacks that are analyzed in detail. This is particularly important for the reader who wants to use a dedicated mean to appreciate whether/how the result of the study can be affected.

This book also offers a short summary of the current areas of research that can benefit of these Earth-bound means. Concerning physical sciences, these areas are fundamental and fluid physics, combustion and materials science. Life science is concerned with cell biology, microbiology, astrobiology, plant sciences, animal/human physiology, radiation, and psychology. Research in ground-based facilities in these disciplines will provide new insights in the fundamental processes of life.

1

The Space Environment

1

The Space Gravity Environment

Daniel A. Beysens[1] and Jack J. W. A. van Loon[2]

[1]CEA-Grenoble and ESPCI-Paris-Tech, Paris, France
[2]VUmc, VU-University, Amsterdam, The Netherlands

It is generally thought that gravity is zero on an object travelling at constant velocity in space. This is not exactly so. We detail in the following those causes that make space gravity not strictly zero.

1.1 Open Space

An object (spacecraft or celestial bodies in general) travelling in the open space is obviously subjected to acceleration forces coming from the spacecraft itself (see below in Section 1.2). The spacecraft is also submitted to gravity forces resulting from the other massive objects, planets, stars, etc. To give an example, Earth's gravity is reduced by a factor 10^6 at a distance of 6×10^6 km from Earth. The Sun's gravity is reduced by the same amount at a distance of 3.7×10^9 km [1].

In addition to these gravity effects, at least in the solar system, a phenomenon of friction due to the solar wind and radiation pressure induces deceleration in the direction of Sun. Basically, absolutely zero gravity does not exist within the universe. All solar systems, nebulas, and galaxies are all under the influence of gravitational fields generated by the mass present and acting over astronomical distances.

1.2 Satellites and Rockets

The means to go into space, that is, going into weightlessness, are classically (sounding) rockets, which follow a parabolic trajectory to generate a free fall or for low-orbit satellites going around Earth where the centrifugal force compensates the gravity attraction. Such spacecrafts are submitted to the above open space effects (Section 1.1). However, there are also effects due to gravity interactions between objects. A mass of 1,000 kg generates 0.007 μg at a 1 m distance. At low orbital altitudes from 185 to 1,000 km, for example, for the International Space Station (ISS), friction effects due to the very sparse molecules in the thermo- and exosphere orbiting the day or night part of an Earth's orbit influence the ISS (Figure 1.1). This atmosphere causes

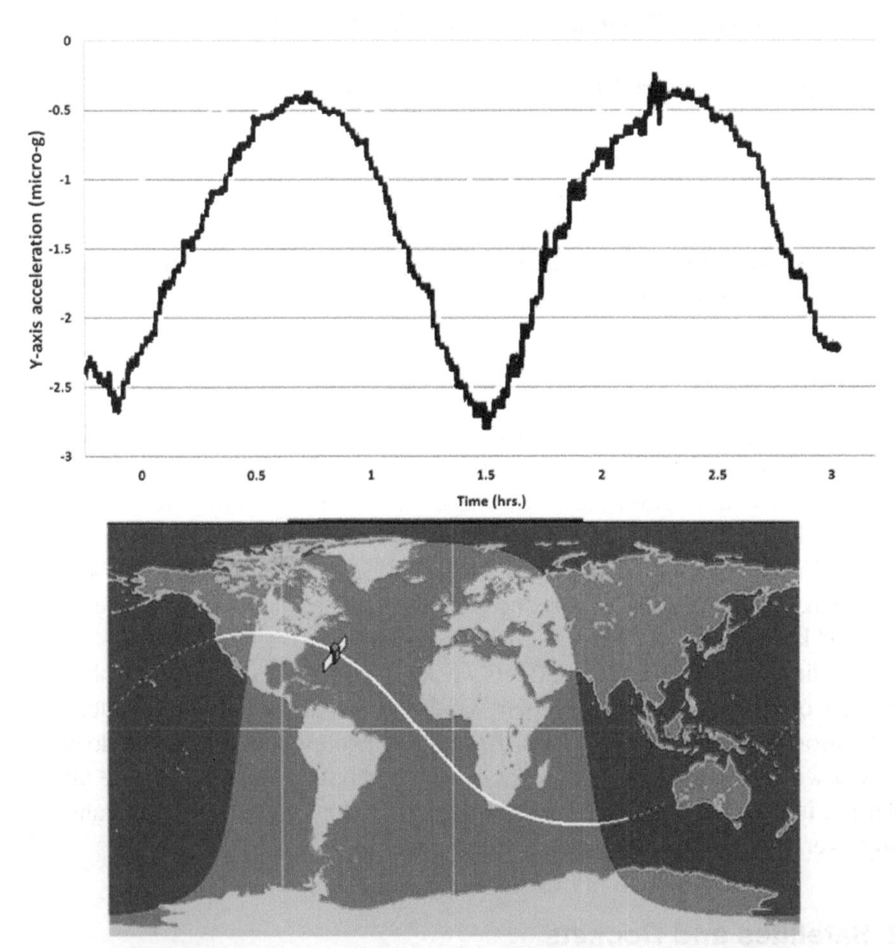

Figure 1.1 The quasi-steady microgravity environment on the orbiter Columbia shows the effects of variations in Earth's atmospheric density. The primary contribution to the variation is the day/night difference in atmospheric density. The *plot* shows that the drag on the orbiter varies over a ninety-minute orbit (Courtesy NASA, 1997 [2]).

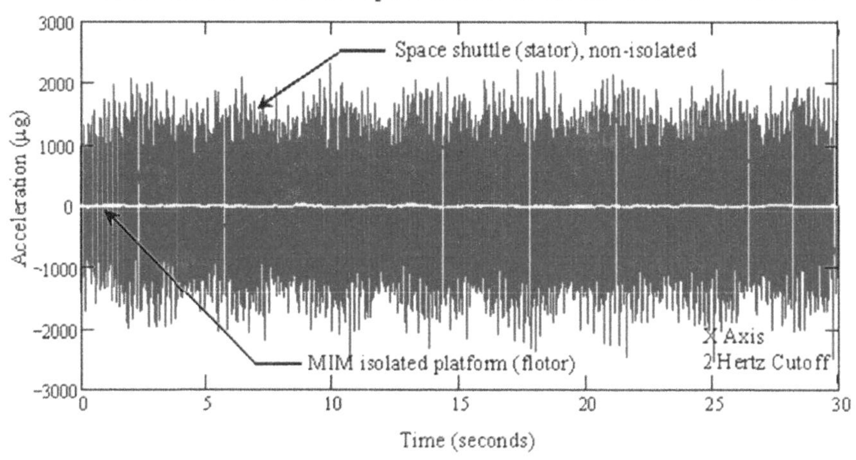

Figure 1.2 Example of the accelerations measured by the accelerometers of the 2nd version of the Microgravity Vibration Isolation Mount (MIM-2) stator (non-isolated) and of the MIM-2 flotor (isolated) during the STS-85 shuttle mission. The accelerations were filtered by a 100-Hz low-pass filter and sampled at 1,000 samples per second. The time traces thus have frequency content up to 100 Hz. The MIM-2 controller was set to isolate above a cutoff frequency of 2 Hz for this run (data courtesy CSA). See also [3].

deceleration, which can be compensated by a small continuous thrust, but in practice, the deceleration is only compensated from time to time, so the small g-force of this effect is not eliminated.

Other effects increase the level of gravity. Even at rest, all the spacecrafts are subjected to vibrations coming from the inboard instrumentation, re-ignition of thrusters, human movement within the module, etc (g-jitter). The vibration frequency spectrum depends on the mechanical architecture of the spacecraft. For instance, in the ISS, the frequency spectrum of vibration in the 10^{-2}–10^{3} Hz range varies from 1 to 10^{3} μ*g* ($g = 9.81\ \mathrm{ms}^{-2}$ is Earth's acceleration constant).

In satellites, the weight compensation is located only at the center of mass. At a distance d from it, gravity rises in the ratio d/z, where z is the distance between the spacecraft and Earth's center of mass. To give an example, at $d = 1$ m from the spacecraft mass center, the residual gravity is steady and equal to 0.17 μ*g*. In addition, in low Earth orbit, the force of gravity decreases upward (by 0.33 μg/m), which can make a variation of about 0.5 μ*g*/m.

Free floating objects in the spacecraft can also be submitted to gravity effects. Such objects orbit Earth in different orbital planes and their distance

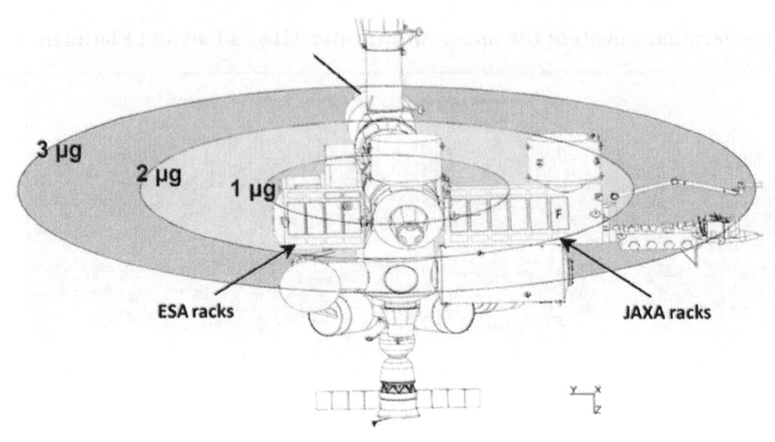

Figure 1.3 Self-gravitation in the International Space Station (*ISS*). Depending on the location with the ISS mass perceives a gravitational pull ranging from 0 to 3×10^{-6} g from the total mass of the ISS itself [4] (Image courtesy NASA).

oscillates, with the same period as the orbit, corresponding to an inward acceleration of 0.17 μg/m. One has also to consider the effects of the solar wind and radiation pressure whose effect is similar to air, but directed away from Sun.

1.3 Typical Gravity at Some Celestial Objects

Gravity amplitude at the surface of some, nearby, celestial objects as asteroids, moons, and planets are listed in Table 1.1, together with their radii. See also Chapter 5 for more information.

Table 1.1 Gravity in units of *g* on some celestial objects

Name	Radius (km)	Gravity (in *g*)
Earth	6,370	1.0
Moon	1,740	0.165
Mars	3,396	0.371
Europa	1,561	0.134
Callisto	2,410	0.126
Io	1,822	0.183
Enceladus	252	0.011
433 Eros[a]	34–11	6.0×10^{-4}
ISS	~0.07 × 0.10	$0–3 \times 10^{-6}$[b]

[a]433 Eros is a near Earth stony asteroid by some mentioned as a possible source for asteroid mining. Estimated mass of the object is 6.7×10^{15} kg (Wikipedia)
[b]Mass of the ISS is around 420,000 kg (date: June 2014)

1.4 Conclusion

The situation of pure weightlessness is thus never encountered in space. What is rather met is a mean low gravity showing at best values expressed in ppm (microgravity), with gravity peaks due to spacecraft maneuvers and human activity. Over the years, various vibration isolation systems have been developed, improving the level of weightlessness, with, however, limitations on (long) times and (small) frequency.

For a typical low Earth orbit, like for the ISS, the altitude is some 350 km. However, the level of gravity is still 9.04 m/s^2 at this height. This is only 8 % less than the gravitational field on Earth's surface.

In this publication, as in many others, the term "microgravity" (μg) is used. Strictly speaking, this term is wrong since, seen the example above, objects are still in a gravitational field, but are in free fall. It is better to use the term weightlessness, or even better, near weightlessness. However, seen the broad use of the term microgravity, we will also apply it in this book.

References

[1] http://en.wikipedia.org/wiki/Micro-g_environment

[2] Microgravity—A Teacher's Guide with Activities in Science Mathematics and Technology EG-1997-08-110-HQ.

[3] http://www.asc-csa.gc.ca/eng/sciences/mimtech.asp#Mir

[4] ISS Design Analysis Cycle & Environment Predictions: Section18. NASA Glenn Research Center, March 5–7, 2002.

2

Cosmos: Violent and Hostile Environment

Sébastien Rouquette

CNES—CADMOS, Toulouse, France

2.1 Introduction

Space attracts, worries, and questions human being for ages, as a frontier set between us and farther horizons to discover. Successive steps were made to unveil its mysteries, before the exploration really started fifty years ago. At the dawn of twentieth century, its main characteristics were not known and we had to wait until the early fifties to get the first data, with the help of balloons, sounding rockets, satellites, and finally human spaceflight to have a better view on this impalpable and elusive place.

2.2 Beliefs and Truths

Scientific description of space started in the middle of the second millennium, at a period when Copernic description of the solar system began to be accepted. Space was then seen as a medium where gods, stars, and comets reside. In Europe, it was described as successive spheres that symbolize way to paradise.

But, after Galileo Galilei (1564–1642), Tycho Brahe (1546–1601), and Copernicus (1473–1543) particularly, scientists started to determine whether space between bodies was empty or filled with a certain invisible fluid. René Descartes (1596–1650) thought that planetary motion was due to *ether* hurly-burly. Christian Huygens (1629–1695), then James Clerk Maxwell (1831–1879), suggested that light and other electromagnetic waves propagate within "light ether." At the beginning of the twentieth century, *ether* hypothesis vanished with the help of Einstein's theory.

By now, cosmologists have to deal with another puzzling story about dark matter and dark energy, a kind of a new step. But our subject is closer to Earth for now...

In the early fifties, some balloons and sounding rockets started to be launched at high latitude in order to study both the edge of the atmosphere and polar aurora phenomenon [1]. It brought the first data on the composition of upper atmosphere, ionosphere, and magnetosphere.

Then, the first human spaceflights sent men out of the protective shelter of Earth [1–4]. Again, these events were a way to answer a variety of questions. Particularly, it was believed that the space environment and weightlessness could deeply affect health and cause some irreversible damages to brain. Gagarin flight was not only a very unique event in a technical point of view, but also for the knowledge it brought on the brain behavior and adaptation in space.

We could now have the slight feeling that everything has been discovered on space. But many challenges are still facing us. We do not know how to safely send a crew to Mars. History of space exploration learnt that the cosmos is a violent and hostile environment!

2.3 Where Space Begins

Earth is wrapped up in a protective shell made of atmosphere and magnetosphere. The atmosphere protects us from neutral particles and high-energy light waves. The magnetosphere extends at a long distance from Earth and acts as a shield against charged particles from solar wind and cosmic rays.

The Kármán line lies at an altitude of 100 km and commonly represents the boundary between Earth's atmosphere and outer space. This definition was endorsed by the Fédération Aéronautique Internationale (FAI), which is a standard setting and record-keeping international entity for aeronautics and astronautics.

At higher altitude, satellites can travel in radiation belts or out of the magnetosphere. This subject is detailed in other sections.

2.4 Satellite Environment

Figure 2.1 shows vertical variation of temperature and pressure from sea level to 900 km. Above 100 km, the atmosphere becomes too thin to support aeronautical flight. There is also an abrupt increase in atmospheric temperature and interaction with solar wind.

Even if the density is lower than 10^{-5} kg/m^3 at satellite altitude, space is not a complete vacuum. It has four major consequences that affect lifetime of satellites. See [5] for further details.

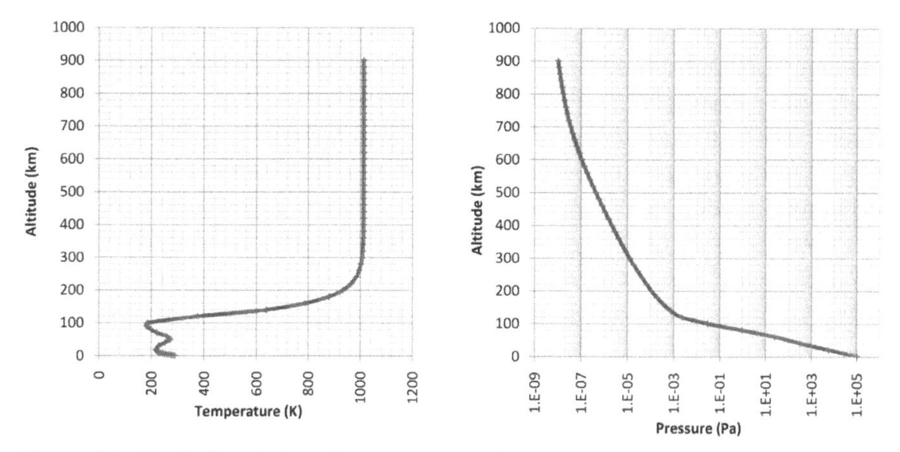

Figure 2.1 Vertical variation of temperature and pressure in Earth's atmosphere (model MSISE-90).

2.4.1 Temperature

In space, heat exchanges can only occur by conduction or radiation, as convection is not efficient anymore. Particles' temperature in the environment is around 1,000 K. But it has low effect on satellites by conduction due to low density/pressure of the atmosphere. Heating of materials is mainly controlled by radiation. It leads to the surprising thing that objects facing to Sun are heated to temperature close to 180 °C, while their shadowed face can have a temperature as low as –180 °C. It might create strong mechanical constraints on structures.

2.4.2 Atmospheric Drag

Considering high velocity of satellites (several km/s depending on altitude and shape of orbit), the residual neutral atmosphere creates a drag that continuously decreases the altitude of orbit. Ground controllers have to correct it regularly. For example, ISS altitude decreases by 50–100 m each day. The orbit has to be corrected 4–6 times a year.

2.4.3 Outgassing

The third effect is due to pressure reduction that creates outgassing of various materials. Moisture, sealants, lubricants, and adhesives are the most common sources, but even metals and glasses can release gases from cracks or impurities. Outgassing products can condense onto optical elements, thermal radiators, or solar cells and obscure them. Space agencies maintain a list of low-outgassing materials to be used for spacecraft. Method of manufacture and preparation can reduce the level of outgassing significantly. Cleaning surfaces or baking individual components or the entire assembly before use can drive off volatiles.

ESA-NASA's Cassini–Huygens space probe has suffered reduced image quality due to a contaminant that condensed onto the CCD sensor of the narrow-angle camera. It was corrected by repeatedly heating the system to 4 °C.

2.4.4 Atomic Oxygen Oxidation

The fourth consequence is due to high-velocity impacts of oxygen atoms on satellite structures. Atomic oxygen density is significant until 1,000 km, and high-energy collisions increase the oxidation power of atoms. For example, oxidation of silver creates an insulating material. Solar array efficiency can be sharply decreased due to interaction between electrical circuit and oxygen.

2.5 Conclusions

The space environment affects the lifetime of satellites. For all the reasons listed in this section, we can conclude that man-made technology is not simple to bring safely into space.

For 50 years, engineers and space specialists in various domains have developed unique skills in order to make spaceships more resistant to temperature contrasts, low pressure, atomic oxygen collisions, and radiations.

The story is still on the way. Not only propulsion will be the key to send more specialized robotic missions and manned spaceships to Mars and beyond. We have to go further in terms of materials innovation and development to build stronger vessels that will be able to protect experiments and crews and send them back safely to Earth.

References

[1] Collective work. "*50 ans de conquête spatiale*". Ciel et espace, Vol. 1, 1999.

[2] Collective work. "*Les débuts de la recherche spatiale française au temps des fusées sondes*". Institut Français d'histoire de l'espace, e/dite, 2007.

[3] Villain, J., and S. Gracieux. "*50 années d'ère spatiale*". Cépaduès Editions, 2007.
[4] Von Braun, W., and F.I. Ordway. "*Histoire mondiale de l'astronautique*". Paris Match—Larousse, 1966.
[5] Collective work. "*Cours de technologie spatiale, Techniques et technologies des véhicules spatiaux*". Conseil Internationnal de la langue Française, PUF, Vol. 1, Module 3, 2002, pp. 323–371.

3

Radiation, Space Weather

Marco Durante

GSI Helmholtzzentrum für Schwerionenforschung and Technische Universität Darmstadt, Darmstadt, Germany

Space radiation has long been acknowledged as a major showstopper for long-term space missions, especially interplanetary, exploratory-class missions [1]. Space radiation is generally divided into three components: trapped radiation, solar particle events (SPEs), and galactic cosmic radiation (GCR). Trapped radiation is the main source of exposure in low Earth orbit (e.g., on the International Space Station), and SPEs are a cause of great concern because they may potentially cause acute radiation syndromes in unprotected crews. However, this type of radiation is mostly composed of protons at energies below 100–200 MeV, and it is therefore relatively easy to shield with conventional bulk materials. On the other hand, GCR contains high-charge and high-energy (HZE) nuclei.

These particles are very penetrating and have high relative biological effectiveness for several late effects. Therefore, energetic heavy ions from the GCR represent the major source of health risk in long-term manned space missions [2].

Cosmic radiation effects could be studied directly in space. This approach has the advantage of including all other space environment factors (microgravity, stress, vibration, etc.) in the experiment. Several radiobiological studies have been carried out during spaceflights [3], but most of the results gathered thus far have been inconclusive. Several factors contribute to the difficulties in interpretation of charged particle radiation effects from spaceflight experiments. The average dose rate in low Earth orbit, though substantially higher than that on Earth, is still fairly low ($\leq$1 mSv/day). The resultant biological effects are small and are often below the detection threshold of most assays, even for long-term missions on the International Space Station. For this reason, radiobiology experiments in space often include high-dose radiation exposure of the sample preflight or onboard, which further complicates the experiment and its interpretation. Second, flight experiments are expensive, difficult to control, restricted to limited sample sizes, and hard to repeat. Very few radiobiology experiments in flight have been repeated, and in the rare cases where this was possible, results were often not confirmed.

It can be safely stated that most of our current knowledge of the health effects of cosmic radiation exposures has been obtained from ground-based experiments at high-energy accelerators [4]. The evaluation of candidate materials for space radiation shielding as well as the effects on microelectronics is also commonly performed by accelerators [5]. Space radiation experimental programs were run at the BEVALAC at the Lawrence Berkeley National Laboratory in Berkeley, CA, USA; the SIS18 at the GSI Helmholtzzentrum für Schwerionenforschung in Darmstadt, Germany (Figure 3.1) [6]; the HIMAC at the National Institute for Radiological Sciences in Chiba, Japan; and at the NASA Space Radiation Laboratory (NSRL) at the Brookhaven National Laboratory in Upton, NY, USA (Figure 3.2) [7]. Many other experiments relevant to extrapolate space radiation effects were performed in other accelerators around the world.

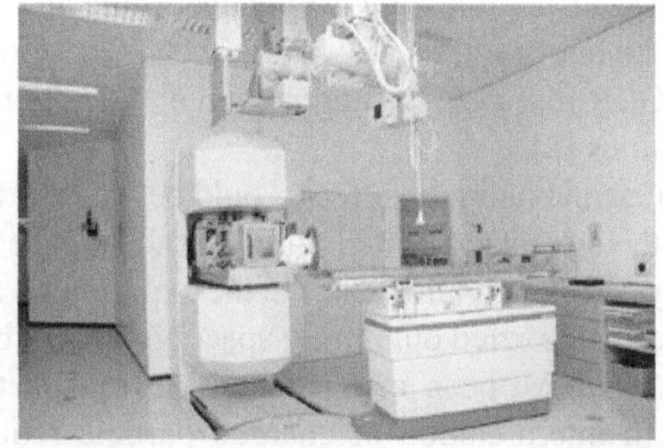

Figure 3.1 Irradiation facility in cave A (left) and cave M (right) from the SIS18 synchrotron of the GSI Helmholtz Center in Darmstadt, Germany. Cave A is equipped with a robotic arm for remote control of the samples. Cave M is equipped with a couch used in 1997–2008 for treatment of cancer patients with C-ions and is currently dedicated to experiments in animals or other 3D targets. Image from the GSI Web site [6].

Figure 3.2 Irradiation facility at NSRL in Upton, NY, USA. The facility is dedicated to the NASA Space Radiation Health Program, the largest research program in the field of simulation of cosmic radiation effects. The photograph showing three large monitor chambers, a plastic target, the egg chamber used for dose measurements, and the digital beam analyzer to check beam position and uniformity. Image from the NSRL Web site [7].

3.1 Facilities for Space Radiation Simulation

Protons are by far the most abundant component in the space radiation environment (see later). In addition, secondary neutrons are produced in space and they can contribute an important fraction of the equivalent dose in shielded areas. Finally, HZE at very high energy can only be properly simulated in large-scale accelerator facilities. It is important to stress that having an accelerator and a cave is not enough to define a "facility" for space radiation research. Specialized infrastructure is necessary—from target handling to beam dosimetry, and including large and expensive tissue culture and animal laboratories for radiobiology experiments.

Accelerators can hardly reproduce the complex space radiation field in space. Generally, only one particle at one defined energy is accelerated.

Often protons around 200 MeV (typical in trapped radiation or SPE) or Fe at 1 GeV/n (representative of the HZE component in the GCR) are used. Moreover, experiments are generally conducted at high dose rate (around 1 Gy/min) and relatively high doses (>0.1 Gy) in most biology experiments. At NSRL (Figure 3.2), an SPE simulator is available that provides protons at different energies simulating energy spectra of past, very intense SPE.

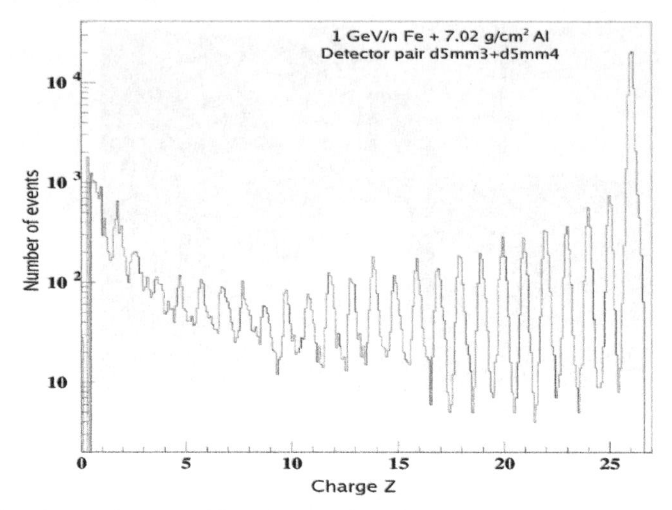

Figure 3.3 Measured spectra of fragments produced by a beam of 1 GeV/n Fe-ions on a target of 26 mm Al. Fragments at 0 degree are measured by a Si-telescope. Each peak correspond to a fragment of a different atomic number Z. Measurement performed at the Brookhaven National Laboratory (NY, USA), courtesy of Jack Miller and Cary Zeitlin, Lawrence Berkeley National Laboratory, CA, USA.

An incubator can also be used on the beamline, thus allowing low-dose rate exposures, even though these experiments are obviously expensive because they burn extended beamtime. Finally, the new electron beam ion source (EBIS) at NSRL allows fast switching between different species and therefore a realistic simulation of the GCR spectrum. Fast energy change is also possible at GSI (Figure 3.1), where a beamline microscope for live microscopy is also available. It should also be noted that a simple simulation of a GCR-like spectrum can be easily obtained using shielding. Using very heavy ions at high energy on a thick target, a fragmentation spectrum is produced which can be modified changing projectile mass and velocity or target material and thickness. An example is shown in Figure 3.3.

A comprehensive review of the facilities for space radiation research was published by the IBER Study Group and supported by the European Space Agency in 2006 [8]. Here, updated information on this topic is summarized.

3.2 Protons

Virtually all electrostatic accelerators and cyclotrons can produce protons at energies below 30 MeV. Moving to the range of 150–250 MeV, there are a number of facilities specialized for proton therapy in oncology which can be used for space radiation research as well. In fact, this is the energy range typical for trapped radiation and SPEs.

Protons are nowadays widely used in cancer treatment [9], thanks to their favorable depth–dose distribution compared to X-rays which can lead to a reduced exposure of the normal tissue with the same conformal coverage of the target. Protons are often used for pediatric tumors, prostate cancers, and head-and-neck tumors. Recently, their use is rapidly increasing also for breast and lung cancers. In June 2014, a total of 105,743 patients had been treated with protons in different facilities in USA, Europe, and Asia [10]. Proton therapy facilities are perfectly equipped with beam delivery and dosimetry systems and often have biology laboratories available, used for preclinical studies. However, availability of beamtime for research is sometimes difficult, because the facilities are very busy treating patients. Currently, very little space-related research has been conducted in clinical proton therapy centers, with the possible exception of Loma Linda in California, where several NASA-supported experiments have been completed. The geographical distribution of the current facilities and those planned or under construction is given in Table 3.1. Facilities for simulation of galactic protons require high-energy machines described in Section 3.4.

3.3 Neutrons

While thermal and fission spectrum neutrons have been studied for many years for radiation protection on Earth, fast neutrons are less characterized and the facilities able to provide reference quasi-monoenergetic neutron fields at energies >20 MeV are only a few.

A recent EURADOS report [11] identified six quasi-monoenergetic neutron facilities in operation worldwide (Table 3.2). These operate in less-than-optimal conditions, especially when seen from the viewpoint of dosimetry. All six facilities make use of the 7Li(p,n) reaction for neutron production. The resulting neutron energy distributions consist of a peak close to the energy of the incoming proton and a broad and roughly even distribution down to zero energy. Each of these components generally contains about half the neutron intensity. A new facility (Neutron for Science (NFS)) is currently under construction in GANIL, France, and is expected to produce quasi-monoenergetic high-energy neutrons from 2014.

Table 3.1 Medical facilities for deep proton therapy (energy >200 MeV) worldwide in operation (at April 2015) and planned or under construction [10]

Location	In Operation	Planned
USA	15	13
Europe	9	16
Russia	3	1
Japan	6	3
China	2	2
South Africa	1	–
Taiwan	–	1
Saudi Arabia	–	1
Australia	–	1

Table 3.2 High-energy quasi-monoenergetic neutron facilities in operation [11]

Name	Country	Energy Range (MeV)
iThemba	South Africa	35–197
TSL	Sweden	11–175
TIARA	Japan	40–90
CYRIC	Japan	14–80
RCNP	Japan	100–400
NPI	Czech Republic	18–36

3.4 Heavy Ions

For the simulation of protons and HZE ions in the GCR, large accelerator facilities are necessary. These facilities are generally synchrotrons, and their main use is either nuclear physics or heavy ion therapy. We will only consider here accelerator facilities capable of providing HZE ions at energies >200 MeV/n. Iron ions are often chosen by space radiation investigators because they are the most abundant specie among the HZE nuclei. The contribution in dose equivalent of Fe alone in deep space is comparable to that of protons.

At March 2013, three facilities deliver both protons and carbon ions in the energy range 200–400 MeV/n for cancer therapy: HIT (Heidelberg, Germany), CNAO (Pavia, Italy), and HIBMC (Hyogo, Japan). These centers, however, are not presently involved in space radiation experiments, even though they have the capability to run this program. On the other hand, the National Institute for Radiological Sciences in Chiba (Japan) and the Institute of Modern Physics of the Chinese Academy of Sciences in Lanzhou (China) treat patients with deep tumors using C-ions and also run extensive space radiation research programs. Both facilities can deliver Fe-ions at energies around 500 MeV/n, and have strong local research groups dedicated to space radiation biology and physics research.

The main research facilities involved in high-energy cosmic ray simulation experiments are NSRL in USA (Figure 3.2) and GSI in Germany (Figure 3.1). The maximum energy available in these facilities is 1–2 GeV/n, depending on the particle mass. Ions up to Au and U have been accelerated at NSRL and GSI, respectively. Most of the space radiation simulation experiments in these facilities, supported by NASA (Space Radiation Health Program) or ESA (IBER program), were however performed with Fe 1 GeV/n. Some space-related studies were also performed at the RIKEN cyclotron in Japan (Z 6, E 135 MeV/n) and at the Joint Institute for Nuclear Research in Dubna, Russia (mass up to Fe, and maximum energy for protons around 6 GeV).

3.5 Facilities Planned

In addition to the new proton therapy centers planned or under construction (Table 3.1), a few new medical centers designed to treat cancer patients with heavy ions are planned or under construction (e.g., MedAustron in Austria: SAGA-HIMAT in Japan, and the Shanghai Proton and Heavy Ion Therapy Hospital in China). However, as noted above, it is unclear how much beamtime can be allocated in these medical facilities to space research.

At least two research facilities are planned in Europe where space radiation research is part of the plans. GSI is now building the facility and antiproton and ion source (FAIR), a double synchrotron with magnetic rigidities of 100 and 300 Tm which will use the current SIS18 as injector [12]. FAIR, which should start operations in 2018, is planning extensive cosmic radiation simulated, extending the ESA support to the current GSI facility. CERN (Geneva, Switzerland) is also considering an experimental biomedical facility based at the low energy ion ring (LEIR) accelerator [13]. Such a new facility could provide beams of light ions (from protons to neon ions) for both cancer therapy and space radiation research projects.

3.6 Conclusions

Ground-based space radiation simulation facilities require large accelerators and dedicated infrastructures. The two main research programs in the field are run at NSRL (NY, USA) with NASA support and GSI (Darmstadt,

Germany) with ESA support. Both facilities have limitations in beamtime and suffer by financial problems in the laboratories (possible RHIC shutdown in Brookhaven; the construction of FAIR in Darmstadt). Particle therapy centers could be used, but access to beamtime is limited in facilities dedicated to patient treatment. New facilities (such as FAIR, NFS, and LEIR in Europe) or more beamtime at therapy centers is needed to run physics, electronics, and biology experiments relevant for space exploration, that is, energetic protons, neutrons, and heavy ions.

References

[1] Radiation Hazards to Crews of Interplanetary Missions. Washington, D.C.: NationalAcademyPress, 1996.
[2] Durante, M., and F.A. Cucinotta, "Heavy Ion Carcinogenesis and Human Space Exploration." *Nature Reviews Cancer* 8, no. 6 (2008): 465–472.
[3] Horneck, G. "Impact of Microgravity on Radiobiological Processes and Efficiency of DNA Repair." *Mutation Research* 430, (1999): 221–222
[4] Durante, M., and A. Kronenberg. "Ground-Based Research with Heavy Ions for Space Radiation Protection." *Advances in Space Research* 35, no. 2 (2005): 180/184.
[5] Durante, M., and F.A. Cucinotta. "Physical Basis of Radiation Protection in Space Travel." *Reviews of Modern Physics* 83, no.4 (2011): 1245–1281.
[6] GSI webpage-http://www.gsi.de/biophysik
[7] NSRL webpage-http://www.bnl.gov/medical/NASA/
[8] Investigation on Biological Effects of Radiation (IBER). Final report. ESA Publication CR(P)-4585, 2006.
[9] Loeffler, J.S., and M. Durante, "Charged Particle Therapy-Optimization, Challenges and Future Directions." *Nature Reviews Clinical Oncology* 10, no.7 (2013): 411–424.
[10] PTCOG webpage-http://http://ptcog.web.psi.ch/
[11] High-Energy Quasi-Monoenergetic Neutron Fields: Existing Facilities and Futureneeds. EURADOS Report 2013–02, Braunschweig, May 2013.
[12] FAIR webpage–http://www.fair-center.eu/
[13] Abler, D., A. Garonna, C. Carli, M. Dosanjh, and K. Peach. "Feasibility Study for a Biomedical Experimental Facility Based on LEIR at CERN." *Journal of Radiation Research* 54, Suppl. 1 (2013): i162–i167.

4

Interstellar Chemistry

Pascale Ehrenfreund

Leiden Observatory, Leiden, The Netherlands

The space between the stars, called the interstellar medium (ISM), is composed primarily of H and He gases incorporating a small percentage of small micron-sized particles. Interstellar clouds constitute a few per cent of galactic mass and are enriched by material ejected from evolved dying stars. Astronomical observations of interstellar clouds have shown dust and molecules widespread in our Milky Way galaxy, as well as distant galaxies [1–3]. The fundamental cloud parameters such as temperature and density can vary substantially. Two main types of interstellar clouds drive molecular synthesis. Cold dark clouds are characterized by very low temperatures ($\sim$10 K) leading to a freeze out of practically all species (except H_2 and He). The higher density of dark clouds ($\sim 10^6$ atoms cm^{-3}) attenuates UV radiation offering an environment where molecules can efficiently form through gas-phase and surface reactions. Surface catalysis on solid interstellar particles enables molecule formation and chemical pathways that cannot proceed in the gas phase owing to reaction barriers [4]. Many molecules have been identified through infrared spectroscopy in ice mantles covering small interstellar dust particles (see Gibb et al. 2004 for a review [5]). Dominated by H_2O, those ice mantles also contain substantial amounts of CO_2, CO and CH_3OH, with smaller admixtures of CH_4, NH_3, H_2CO and HCOOH [6, 7]. The median ice composition H_2O:CO:CO_2:CH_3OH:NH_3:CH_4:XCN is 100:29:29:3:5:5:0.3 and 100:13:13:4:5:2:0.6 towards low- and high-mass protostars, respectively, and 100:31:38:4:-:-:- in cloud cores [8]. Laboratory simulations indicate that thermal and UV radiation processing close to the protostars results in ice desorption, ice segregation and the formation of complex organic molecules such as quinones and even dipeptides [9, 10].

Diffuse interstellar clouds are characterized by low densities ($\sim 10^3$ atoms cm^{-3}) and temperatures of $\sim$100 K. Ion–molecule reactions, dissociative recombination with electrons, radiative association reactions and neutral–neutral reactions contribute to gas-phase processes and influence molecule formation in those regions. Many small molecules, including CO, CH, CN, OH, C_2, C_3 and C_3H_2, have been observed [11, 12], together with a high fraction of polycyclic aromatic hydrocarbons (PAHs) [2]. Dust particles contain macromolecular aromatic networks evidenced by a ubiquitous strong UV absorption band at 2175 Å [13]. Amorphous carbon, hydrogenated amorphous carbon, diamonds, refractory organics and carbonaceous networks such as coal, soot and graphite and quenched carbonaceous condensates have been proposed as possible carbon compounds [14, 15]. In diffuse interstellar clouds, dust interacts with hot gas, UV radiation and cosmic rays, and evolves or gets destroyed in shocks and by sputtering. Strong differences in the dust component of dense and diffuse interstellar clouds exclude rapid cycling of cloud material [16].

In summary, a large number of complex molecules in the gas phase have been identified through infrared, radio, millimetre and sub-millimetre observations. Currently, > 180 molecules are detected in the interstellar and circumstellar gas although some of them are only tentatively identified and need confirmation. More than 50 molecules are found in extragalactic sources (http://www.astro.uni-koeln.de). H_2 is by far the most abundant molecule in cold interstellar regions, followed by CO, the most abundant carbon-containing species, with $CO/H_2 \sim 10^{-4}$. The chemical variety of molecules includes nitriles, aldehydes, alcohols, acids, ethers, ketones, amines and amides, as well as long-chain hydrocarbons.

Circumstellar envelopes, regarded as the largest factories of carbon chemistry in space, are where small carbon compounds are converted to larger species and into solid aromatic networks such as soot [17]. Processes analogous to soot formation on terrestrial environments are assumed to form robust coal-like material. Laboratory simulations showed that the temperature in the circumstellar condensation zone determines the formation pathway of carbonaceous particles with lower temperatures ($<$1700K) producing PAHs with three to five aromatic rings [18] compared to temperatures above 3500 K that favour the production of fullerene compounds. The detection of C_{60} and C_{70} fullerene molecules was recently reported in a protoplanetary nebula by Cami et al. [19]. Molecular synthesis may occur in the circumstellar environment on timescales as short as a few hundred years.

Interstellar chemistry shapes the raw material for the formation of stars and planets. The gravitational collapse of an interstellar cloud led to the formation of the protosolar nebula approximately 4.6 billion years ago. From this solar nebula, planets and small bodies formed within less than 50 million years. Data from recent space missions, such as the Spitzer telescope, Herschel, Stardust and Deep Impact, show a dynamic environment of the solar nebula with the simultaneous presence of gas, particles and energetic processes, including shock waves, lightning and radiation. The carbonaceous inventory of our solar system has therefore experienced a variety of conditions and contains a mixture of material that was newly formed in the solar nebula as well as interstellar material that experienced high temperatures and radiation. Some pristine cloud material with significant interstellar heritage has survived as evidenced from laboratory studies of extraterrestrial material. Understanding the evolution of interstellar material and dust cycling provides important insights into the nature of the material that is later incorporated into planet and small bodies. The latter, including comets, asteroids and their fragments, carbonaceous meteorites and micrometeorites, contain a variety of molecules including biomarkers that transported raw material for life to the young planets via impacts in the early history of the solar system [20]. Experiments in low Earth orbit enable to simulate true space conditions and have contributed important results on the stability and photochemistry of organic compounds, biomarkers and microbes in space environment in the last decade.

References

[1] Snow, T.P., and B.J. McCall. "Diffuse Atomic and Molecular Clouds." *Annual Review of Astronomy and Astrophysics* 44 (2006): 367–414.

[2] Tielens, A.G.G.M. "Interstellar Polycyclic Aromatic Hydrocarbon Molecules." *Annual Review of Astronomy and Astrophysics* 46 (2008): 289–337.

[3] Harada, N., T.A. Thompson, and E. Herbst. "Modeling the Molecular Composition in an Active Galactic Nucleus Disk." *The Astrophysical Journal* 765 (2013): 26.

[4] Cuppen, H.M., and E. Herbst. "Simulation of the Formation and Morphology of Ice Mantles on Interstellar Grains." *The Astrophysical Journal* 668 (2007): 294–309.

[5] Gibb, E., D. Whittet, A. Boogert, and A.G.G.M. Tielens. "Interstellar Ice: The Infrared Space Observatory Legacy." *The Astrophysical Journal Supplement Series* 151 (2004): 35–73.

[6] Boogert, A., et al. "The c2d Spitzer Spectroscopic Survey of Ices Around Low-Mass Young Stellar Objects. I. H_2O and the 5–8 mm Bands." *The Astrophysical Journal* 678 (2008): 985–1004.

[7] Pontoppidan, K.M., A.C.A. Boogert, H.J. Fraser, E.F. van Dishoeck, G.A. Blake, F. Lahuis, K.I. Öberg, N.J. Evans II, and C. Salyk. "The c2d Spitzer Spectroscopic Survey of Ices Around Low-Mass Young Stellar Objects. II. CO_2." *The Astrophysical Journal* 678 (2008): 1005–1031.

[8] Öberg, K., et al. "The Spitzer Ice Legacy: Ice Evolution from cores to protostars." *The Astrophysical Journal* 740 (2011): 16 pp.

[9] Bernstein, M.P., S.A. Sandford, L.J. Allamandola, J.S. Gillette, S.J. Clemett, and R.N. Zare. "UV Irradiation of Polycyclic Aromatic Hydrocarbons in Ices: Production of Alcohols, Quinones, and Ethers." *Science* 283 (1999): 1135–1138.

[10] Kaiser, R.I., A. Stockton, Y. Kim, E. Jensen, and R.A. Mathies. "On the Formation of Dipeptides in Interstellar Model Ices." *The Astrophysical Journal* 765 (2013): 9 pp.

[11] Liszt, H., and R. Lucas. "The Structure and Stability of Interstellar Molecular Absorption Line Profiles at Radio Frequencies." *Astronomy and Astrophysics* 355 (2000): 333–346.

[12] Liszt, H., P. Sonnentrucker, M. Cordiner, and M. Gerin. "The Abundance of C3H2 and Other Small Hydrocarbons in the Diffuse Interstellar Medium." *The Astrophysical Journal* 753 (2013): 5.

[13] Mennella, V., L. Colangeli, E. Bussoletti, P. Palumbo, and A. Rotundi, "A New Approach to the Puzzle of the Ultraviolet Interstellar Extinction Bump." *The Astrophysical Journal* 507 (1998): 177–180.

[14] Henning, T., and F, Salama. "Carbon in the Universe." *Science* 282 (1998): 2204–2210.

[15] Pendleton, Y.J., and L.J. Allamandola. "The Organic Refractory Material in the Diffuse Interstellar Medium: Mid-Infrared Spectroscopic Constraints." *Astrophysical Journal Supplement Series* 138 (2002): 75–98.

[16] Chiar, J.E., and Y. Pendleton. "The Origin and Evolution of Interstellar Organics." *Organic Matter in Space, Proceedings of the International Astronomical Union, IAU Symposium,* vol. 251, pp. 35–44. Cambridge, UK: Cambridge University Press, 2008.

[17] Kwok, S. "Delivery of Complex Organic Compounds from Planetary Nebulae to the Solar System." *The International Journal of Astrobiology* 8 (2009): 161–167.

[18] Jäger, C., F. Huisken, H. Mutschke, I. Llamas Jansa, and Th. Henning. "Formation of Polycyclic Aromatic Hydrocarbons and Carbonaceous

Solids in Gas-Phase Condensation Experiments." *The Astrophysical Journal* 696 (2009): 706–712.

[19] Cami, J., J. Bernard-Salas, E. Peeters, and S.E. Malek. "Detection of C60 and C70 in a Young Planetary Nebula." *Science* 329 (2010): 1180.

[20] Ehrenfreund, P., et al. "Astrophysical and Astrochemical Insights into the Origin of Life." *Reports on Progress in Physics* 65 (2002): 1427–1487.

5

Celestial Bodies

Inge Loes ten Kate[1] and Raheleh Motamedi[2]

[1]Utrecht University, Utrecht, The Netherlands
[2]VU University, Amsterdam, The Netherlands

5.1 Introduction

The previous paragraphs have described several parameters that play a role in space. In this paragraph, we focus on conditions occurring on planetary surfaces that are reproducible in a laboratory setting, which include some of the parameters described earlier. Most simulation facilities are designed to reproduce atmospheric pressure and composition, ultraviolet (UV) radiation, and surface temperature. Table 5.1 gives an overview of the surface conditions on several terrestrial bodies, as well as the Moon and Titan. So far, there are no simulation facilities known that focus on other solar system bodies.

The majority of the simulation facilities focuses on a range of planetary surface conditions either for scientific studies or for instrument testing, and examples of these are described below. Some also focus on more specific scenarios, including wind tunnels that are described below as well.

5.2 General Planetary Simulation Facilities

5.2.1 The Centre for Astrobiology Research (CAB), Madrid, Spain

CAB houses a versatile environmental simulation chamber capable of reproducing atmospheric compositions and surface temperatures for most planetary objects. This 50 $\times$ 40 cm chamber was specifically developed to subject samples to in situ irradiation. The internal pressure can be varied between 5 and 5×10^{-9} mbar. The required atmospheric composition is regulated using a residual gas analyzer with ppm precision. Temperatures can be set from 4 K to 325 K UV radiation is provided by a combination of a deuterium

Table 5.1 Selected surface and atmospheric parameters of selected solar system bodies (adapted from [1])

Solar System Body	Mercury	Venus	Earth
Mass (10^{24} kg)	0.33	4.87	5.97
Radius (km)	2,439.7	6,051.8	6,378.14
Density (g cm^{-3})	5.43	5.24	5.52
Surface gravity (m s^{-2})	3.70	8.87	9.80
Temperature (K)	100–700	737	184–330
Escape velocity (km s^{-1})	4.25	10.36	11.18
Length of day (h)	4,222.6	2,802.0	24.0
Atmospheric pressure (mbar)	10^{-11}	95.6×10^3	1,000
Atmospheric composition	42 % O (molecular)	96.5 % CO_2	78.08 % N_2
	29 % Na	3.5 % N_2	20.95 % O_2
	22 % H	0.015 % SO_2	0.93 % Ar
	6 % He	0.007 % Ar	0.036 % CO_2
	Traces Na, K, Ca, Mg	0.002 % H_2O (vapor)	~1 % H_2O (vapor)
	Mars	Moon	Titan
Mass (10^{24} kg)	0.64	0.07	0.13
Radius (km)	3,396.2	1,738.1	2,575.5
Density (g cm^{-3})	3.93	3.35	1.88
Surface gravity (m s^{-2})	3.71	1.62	1.35
Temperature (K)	130–308	100–390	93.7
Escape velocity (km s^{-1})	5.03	2.38	2.65
Length of day (h)	24.7	708.7	382.7
Atmospheric pressure (mbar)	10	10^{-9}(day)–10^{-12}(night)	1,467
Atmospheric composition	95.32 % CO_2	Ar	82–99 % N_2
	2.7 % N_2	H	1–6 % CH_4
	1.6 % Ar	Na	Traces of
	0.13 % O_2	H	Ar, H_2,
	0.08 % CO	K	C_2H_2, C_2H_4,
	0.03 % H_2O (vapor)		C_2H_6, C_3H_4,
			C_3H_8, C_4H_2,
			HCN, HC_3N,
			C_6H_6, C_2H_2

lamp and a noble gas discharge lamp. The chamber has *in situ* analytical capabilities in the form of UV spectroscopy and infrared spectroscopy (IR). This chamber is especially suitable for following the chemical changes induced in a particular sample by irradiation in a controlled environment. Therefore, it can be used in different disciplines such as planetary geology, astrobiology, environmental chemistry, and materials science as well as for instrumentation testing [2].

5.2.2 Deutsches Zentrum fur Luft-und Raumfahrt (DLR), Berlin, Germany

There are two planetary simulation chambers in the Planetary Emissivity Laboratory at DLR. One is a vacuum chamber (approximately 40 × 30 × 30 cm) and simulates conditions on Venus and Mercury. Samples can reach 773 K and beyond, while keeping the rest of the chamber relatively cold [3]. The second chamber is a Mars simulation facility (MSF). The MSF laboratory consists of a cold chamber with a cooled volume of 80 × 60 × 50 cm. The effective operational experimental chamber, which is cooled within the cold chamber, is a cylinder with inner diameters of 20.1 × 32.4 cm. This chamber operates at 6 mbar CO_2 pressure at 198 K [4].

5.2.3 The Open University, Milton Keynes, UK

This Mars simulation facility consists of a large chamber (90 × 180 cm), providing pressure and temperature conditions representative of the surface conditions on Mars. This chamber is configured with the capability to incorporate large-scale regolith experiments not usually possible within standard vacuum systems. Another chamber is a small Mars chamber (70 × 100 cm) providing a simulated Martian environment with a solar illumination facility designed for instrument qualification and astrobiology experiments. The facility is also configured to permit automated variation of the environment, such as thermal diurnal cycling [5].

5.2.4 Mars Environmental Simulation Chamber (MESCH), Aarhus University, Denmark

MESCH (Figure 5.1) is a dynamic simulation facility, providing low temperature (down to 133 K), low atmospheric pressure (5–10 mbar), and a gas composition like that of Mars during long-term experiments. The main chamber is cylindrical cryogenic environmental chamber, with a double wall providing a cooling mantle through which liquid N_2 can be circulated. The chamber is equipped with an atmospheric gas analyzer and a xenon/mercury discharge source for UV generation. Exchange of samples without changing

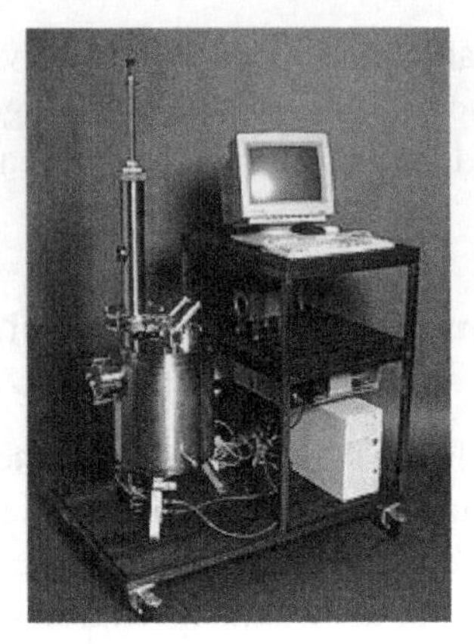

Figure 5.1 The Mars environmental simulation chamber. *Image credit* Mars Simulation Laboratory, Aarhus University.

the chamber environment is possible through a load lock system consisting of a small pressure-exchange chamber that can be evacuated. Within the MESCH, up to 10 steel sample tubes can be placed in a carousel that is controlled by an external motor to allow any desired position. A wide variety of experiments is possible through computer logging of environmental data, such as temperature, pressure, and UV exposure time, and automated feedback mechanisms [6].

5.2.5 The Planetary Analogues Laboratory for Light, Atmosphere and Surface Simulations (PALLAS), Utrecht University, The Netherlands

PALLAS (Figure 5.2) is designed to study organic processes in a planetary surface environment, simulating ultraviolet radiation, surface temperature, humidity, and atmospheric composition. PALLAS is a 50 × 50 × 50 cm stainless steel vacuum chamber equipped with a differentially pumped sampling volume for real-time atmospheric measurements, the atmospheric sample chamber (ASC). The ASC is equipped with a turbo pump attached to a diaphragm pump, a mass spectrometer, and a pressure gauge. A xenon arc discharge lamp provides the desired solar spectrum and irradiates the samples through a UV-transparent fused-silica window. An airtight tube is mounted between the lamp housing and the fused-silica window and can be filled with N_2 to minimize UV loss and ozone formation. Samples are placed on temperature-controlled tables and can variably be irradiated in the beam spot

Figure 5.2 The Planetary Analogues Laboratory for Light, Atmosphere and Surface Simulations, Utrecht University.

of the UV source. The temperature of the sample tables is controlled using a refrigerated heating circulator. Three gas inlet valves are connected to the chamber to insert atmospheric gases. One inlet is connected to a N_2 line, used to vent the chamber while preventing atmospheric water from entering. Gases can be either premixed or mixed inside the chamber to obtain the desired atmospheric conditions. Atmospheric pressures inside the chamber are monitored with a pressure gauge [1].

5.3 Mars Wind Tunnels

Dust devils and dust storms occur on a regular basis on the Martian surface, leading to a range of processes. The abrading effect of dust on landers and rovers is one important process, albeit more more engineering than astrobiology. More interesting from an astrobiological perspective are the effects of dust abrasion on mineralogy [7] and the generation of an electric field by dust interaction, a process detected in terrestrial dust storms (e.g., [8]). Martian dust storms have furthermore been related to oxygen enhancement [9] and methane destruction in the atmosphere [10]. A selection of wind tunnel facilities is described below.

5.3.1 The Planetary Aeolian Laboratory (PAL), NASA Ames Research Center, Moffett Field, CA, USA

PAL (Figure 5.3) is a pentagon-shaped, concrete chamber 30 m high, with a floor area of 164 m^2 and a total chamber volume of 4,058 m^3. The entire chamber can be evacuated to a minimum pressure of 3.8 mbar. A 7.6 m × 7.9 m

Figure 5.3 The Planetary Aeolian Laboratory. *Image credit* NASA Ames Research Center.

door permits large experimental apparatus to be placed inside the chamber. PAL contains three separate facilities: the Venus Wind Tunnel [11, 12], the Arizona State University Vortex Generator (see below), and the *Mars Surface Wind Tunnel (MARSWIT)*, the first Mars wind tunnel, established in the 1960s at NASA Ames Research Center [13–16]. MARSWIT occupies the center of the PAL with an overall length of 14 m. The tunnel walls are constructed of 2.4-cm-thick clear Plexiglas to enabling ready viewing, and a 1.1-m^2 test section is located 5 m from the entrance. The tunnel is driven by a network ejector system consisting of 72 equally spaced 1.6-m nozzles located in the diffuser section. High-pressure air (up to 9.86 kg/cm^2) is forced through the nozzles to induce flow of air through the tunnel. The maximum attainable free stream airspeed is 13 m/s at atmospheric pressure, increasing to 180 m/s at 5 mbar (500 Pa) [17].

5.3.2 The Arizona State University Vortex Generator (ASUVG), Moffett Field, CA, USA

The ASUVG was built to simulate dust devils in the laboratory and consists of three components. The *vortex generator* includes a cylinder (45 cm in diameter by 1.3 m long) with a "bell mouth" to alleviate boundary effects at the edge of the cylinder, a motor drive, and a fan blade system. To vary the geometry of the simulated dust devil, the generator is mounted to *a frame* so that it can be lowered or raised above the test table. The *table* is 2.4 by

2.4 m, mounted independent of the frame so that potential motor vibrations are isolated from the test bed. The table can be raised or lowered, moved laterally to simulate motion of a dust devil across terrain features, and tilted to simulate a vortex that is not perpendicular to the surface. The facility is equipped with instruments enabling real-time measurements of the ambient temperature, relative humidity, and wind speeds and surface pressures on the test bed beneath the vortex. The generator can be dismantled for, for example, transport into the field for conducting experiments on natural surfaces and use in the MARSWIT for tests under Martian atmospheric conditions [15, 16, 18, 19].

5.3.3 The Aarhus Wind Tunnel Simulator (AWTS), Aarhus, Denmark

The AWTS (Figure 5.4) consists of a recirculating wind tunnel housed inside an environmental chamber and is designed to reproduce the environmental conditions observed at the surface of Mars, specifically the atmospheric pressure and composition, the temperature, wind conditions, and the transport of airborne dust. The environmental chamber is 0.8 m wide and 3 m long and can be evacuated to around 0.03 mbar and repressurized and held at Mars-like pressures (typically 6–10 mbar). The central wind tunnel is cylindrical, 0.4 m in diameter, and 1.5 m long. To maximize the available open wind tunnel area (cross section) while maintaining smooth fluid flow, an axially mounted fan driven by an electric motor draws gas down the central wind tunnel and returns it in an outer cylindrical cavity. Mechanical obstructions that may create excessive turbulence are avoided, and smooth surfaces are used wherever possible [7, 20–24].

Figure 5.4 The Aarhus wind tunnel simulator. *Image credit* Mars Simulation Laboratory, Aarhus University.

5.4 Instrument Testing Facilities

To test planetary instruments under realistic conditions, several chambers have been developed specifically for instrument testing. These facilities mimic planetary conditions as well, but are in general not used for science, but solely for instrument testing.

5.4.1 ChemCam Environmental Chamber

The ChemCam environmental chamber was developed at the University of Toulouse, France, to reproduce the Martian environment to test the first laser-induced breakdown spectroscopy instrument sent into space as part of ChemCam on the Curiosity Rover [25]. The chamber has a volume of 70 l. The chamber is pumped to 10^{-3} mbar and then filled with 95.7 % CO_2, 2.7 % N_2, and 1.6 % Ar to mimic the Martian atmosphere. In each experiment, five samples are placed in the chamber and the ChemCam instrument is installed 3 m from the sample. The chamber is kept at room temperature, which is a difference compared to flight mode conditions, but this should be generally of no importance for laser-induced breakdown spectroscopy (LIBS) analysis because of the high temperature of the plasma ~8,000 °C [26].

5.4.2 SAM Environmental Chamber

The Sample Analysis at Mars instrument suite (SAM) environmental chamber (Figure 5.5) was developed at NASA Goddard Space Flight Center to carry out both thermal testing and qualification and calibration in an environment that could simulate the thermal conditions in the rover on the surface of Mars. The design of the chamber enables simultaneous instrument suite qualification, through thermal cycling, and calibration, utilizing both solid samples and atmospheric samples introduced into SAM through chamber feedthroughs from a gas processing system external to the chamber. The SAM chamber is a ~91 cm electro-polished stainless steel cube fitted with an internal thermal shroud, with an internal test volume of 66 × 56 × 41 cm. This volume consists of six independent thermal zones, where the temperature can be cycled between 233 and 323 K. The pressure in the chamber can be varied from 10^{-6} to 1,000 mbar. The Mars chamber is equipped with a dedicated 120-channel thermocouple data acquisition system and standard contamination control and monitoring systems to include a thermoelectric quartz crystal microbalance (TQCM), residual gas analyzer (RGA) and scavenger plate [27].

Figure 5.5 The SAM environmental chamber. *Image credit* NASA Goddard Space Flight Center.

References

[1] ten Kate, I.L. and M. Reuver. "PALLAS: Planetary Analogues Laboratory for Light, Atmosphere, and Surface Simulations". *Netherlands Journal of Geosciences* (2015, in press).

[2] Mateo-Martí, E., O. Prieto-Ballesteros, J.M. Sobrado, J. Gómez-Elvira and J.A. Martín-Gago. "A chamber for studying planetary environments and its applications to astrobiology". *Measurement Science and Technology* 17, no. 8 (2006): 2274–2280.

[3] Maturilli, A., J. Helbert and M. D'Amore. "Dehydration of Phyllosilicates under Low Temperatures: An Application to Mars". In *41st Lunar and Planetary Science Conference*, abstract 1533, 2010.

[4] De Vera, J.P., D. Möhlmann, F. Butina, A. Lorek, R. Wernecke and S. Ott. "Survival Potential and Photosynthetic Activity of Lichens under Mars-Like Conditions: A Laboratory Study". *Astrobiology* 10, no. 2 (2010): 215–227.

[5] Patel, M.R., K. Miljkovic, T.J. Ringrose and M. R. Leese. "The Hypervelocity Impact Facility and Environmental Simulation at the Open University". In *5th European Planetary Science Congress*, 655, 2010.

[6] Jensen, L.L., J. Merrison, A.A. Hansen, K.A. Mikkelsen, T. Kristoffersen, P. Nørnberg, B.A. Lomstein and K. Finster. "A Facility for Long-Term Mars Simulation Experiments: The Mars Environmental Simulation Chamber (MESCH)". *Astrobiology* 8 (2008): 537–548.

[7] Merrison, J.P., H.P. Gunnlaugsson, S. Knak Jensen and P. Nørnberg. "Mineral Alteration Induced by Sand Transport: A Source for the Reddish Color of Martian Dust". *Icarus* 205, no. 2 (2010): 716–718.

[8] G. Freier. "The Electric Field of a Large Dust Devil". *Journal of Geophysical Research* 65, no. 10 (1960): 3504–3504.

[9] Atreya, S.K., A. Wong, N. Renno, W. Farrell, G. Delory, D. Sentman, S. Cummer, J. Marshall, S. Rafkin and D. Catling. "Oxidant Enhancement in Martian Dust Devils and Storms: Implications for Life and Habitability". *Astrobiology* 6, no. 3 (2006): 439–450.

[10] Farrell, W., G. Delory and S. K. Atreya. "Martian Dust Storms as a Possible Sink of Atmospheric Methane". *Geophysical Research Letters* 33, no. 21 (2006): L21203.

[11] Greeley, R., J. Iversen, R. Leach, J. Marshall, B. White and S. Williams. "Windblown Sand on Venus—Preliminary Results of Laboratory Simulations. *Icarus* 57 (1984): 112–124. doi:10.1016/0019-1035(84)90013-7.

[12] Greeley, R., J.R. Marshall and R.N. Leach. "Microdunes and Other Aeolian Bedforms on Venus—Wind Tunnel Simulations". *Icarus* 60 (1984): 152–160.

[13] Greeley, R., R. Leach, B. White, J. Iversen and J. Pollack. "Threshold Windspeeds for Sand on Mars—Wind-Tunnel Simulations". *Geophysical Research Letters* 7, no. 2 (1980): 121–124.

[14] Greeley, R., G. Wilson, R. Coquilla, B. White and R. Haberle. "Windblown Dust on Mars: Laboratory Simulations of Flux as a Function of Surface Roughness". *Planetary and Space Science* 48, no. 12–14 (2000): 1349–1355.

[15] Greeley, R. "Saltation Impact as a Means for Raising Dust on Mars". *Planetary and Space Science* 50, no. 2 (2002): 151–155.

[16] Greeley, R., M.R. Balme, J.D. Iversen, M. Metzger, R. Mickelson, J. Phoreman and B. White. "Martian Dust Devils: Laboratory Simulations of Particle Threshold". *Journal of Geophysical Research* 108 (2003): 5041.

[17] Greeley, R., B.R. White, J.B. Pollack, J.D. Iversen and R.N. Leach. "Dust Storms on Mars: Considerations and Simulations". *NASA Technical Memorandum* 78423 (1977): 1–32.

[18] Neakrase, L.D.V., R. Greeley, J.D. Iversen, M.R. Balme and E.E. Eddlemon. "Dust Flux Within Dust Devils: Preliminary Laboratory Simulations". *Geophysical Research Letters* 33, no. 19 (2006): L19S09.

[19] Neakrase, L.D.V. and R. Greeley. "Dust Devil Sediment Flux on and Mars: Laboratory Simulations". *Icarus* 206, no. 1 (2010): 306–318.

[20] Merrison, J.P., P. Bertelsen, C. Frandsen, P. Gunnlaugsson, J.M. Knudsen, S. Lunt, M.B. Madsen, L.A. Mossin, J. Nielsen, P. Nørnberg, K.R. Rasmussen and E. Uggerhoj. "Simulation of the Martian Dust Aerosol at Low Wind Speeds". *Journal of Geophysical Research—Planets* 107 (2002): 5133.
[21] Merrison, J.P., J. Jensen, K. Kinch, R. Mugford and P. Nørnberg. "The Electrical Properties of Mars Analogue Dust". *Planetary and Space Science* 52, no. 4 (2004): 279–290.
[22] Merrison, J.P., H. Gunnlaugsson, P. Nørnberg, A. Jensen and K. Rasmussen. "Determination of the Wind Induced Detachment Threshold for Granular Material on Mars Using Wind Tunnel Simulations". *Icarus* 191, no. 2 (2007): 568–580.
[23] Merrison, J.P., H. Bechtold, H. Gunnlaugsson, A. Jensen, K. Kinch, P. Nørnberg and K. Rasmussen. "An Environmental Simulation Wind Tunnel for Studying Aeolian Transport on Mars". *Planetary and Space Science* 56 (2008): 426–437.
[24] Merrison, J.P., H.P. Gunnlaugsson, M.R. Hogg, M. Jensen, J.M. Lykke, M. Bo Madsen, M.B. Nielsen, P. Nørnberg, T.A. Ottosen, R.T. Pedersen, S. Pedersen and A.V. Sørensen. "Factors Affecting the Electrification of Wind-Driven Dust Studied with Laboratory Simulations". *Planetary and Space Science* 60, no. 1 (2012): 328–335.
[25] Cousin, A., O. Fornia, S. Maurice, O. Gasnault, C. Fabre, V. Satter, R.C. Wiens and J. Mazoyer. "Laser Induced Breakdown Spectroscopy Library for the Martian Environment". *Spectrochimica Acta Part B: Atomic Spectroscopy* 66, no. 11–12 (2011): 805–814.
[26] Cremers, D. and L. Radziemski. "Hystory and Fundamentals of LIBS". In *Laser Induced Breakdown Spectroscopy: Fundamentals and Applications*. (Cambridge University Press, 2006), 9–16.
[27] Mahaffy, P.R., et al. "The Sample Analysis at Mars Investigation and Instrument Suite". *Space Science Reviews* 170, no. 1–4 (2012): 401–478.

2

Facilities to Alter Weight

6

Drop Towers

Claus Lämmerzahl[1] and Theodore Steinberg[2]

[1]University of Bremen, Bremen, Germany
[2]Queensland University of Technology,
Brisbane, Australia

6.1 Introduction

While the utilization of space-based or flight-based facilities can provide longer-duration test times than can be provided by Earth-based facilities, access to these facilities comes with an associated significant increase in cost and often an associated decrease in availability. These problems are somewhat mitigated by the use of ground-based facilities that are often able to provide very good levels of reduced gravity coupled with low cost (per test) and significantly better access (than flight-based or space-based facilities). This section provides a brief overview of ground-based facilities that are able to provide periods of reduced gravity for the testing of various phenomena in many diverse disciplines. Described in detail are the different types of facilities available and the principle used in these facilities to produce low-gravity conditions. A short description of the various discipline areas currently utilizing these ground-based facilities is included for completeness.

Ground-based methods permit (with low cost, good access, and high test rate) the conduct of complex experiments. Experiments useful in many diverse discipline areas have been, and will continue to be, conducted in the reduced gravity environment produced inside a drop tower. These discipline areas include materials, fluids, astrophysics, phase transitions, combustion, fire safety, fundamental physics, biology and life sciences, heat transfer, mechanics, and technology development.

6.2 Drop Tower Technologies

The first drop test performed, attributed to Galileo (in the late 1500s), supposedly occurred from the Leaning Tower of Pisa to demonstrate that objects fall, independent of their mass, at the same acceleration in the Earth's gravitational field. That is, an object in free fall is essentially in zero gravity. A problem for a falling object in a fluid medium such as air, however, is the development of aerodynamic drag, which results in the slowing down of the falling object, thus reducing (or eliminating) the reduced gravity conditions as the object approaches its terminal velocity. A ground-based facility that is providing reduced gravity conditions must, in some fashion, reduce or eliminate the presence of Earth's gravity and the effects of this body force on the phenomena being investigated while, at the same time, eliminating the detrimental effects of aerodynamic drag. This is accomplished by accelerating the experiment, at 1 g, in a vector parallel to Earth's gravity relative to the Earth's centered frame of reference. Matching the Earth's acceleration essentially produces a free fall environment within which the experiment is in a zero-gravity condition in the freely falling reference frame. The precision to which the experiment's acceleration is matched to Earth's gravity level dictates the quality of the reduced gravity that is obtained.

The drag produced, and its effects, on an object moving through a fluid, has been well studied and characterized over many years, and this work is well documented in publications due to its relevance to many disciplines. Drag coefficients have been developed and allow researchers to predict and plan for the effects of the aerodynamic drag, as required, as a function of an object's geometry and the flow conditions present.

There are several options available to eliminate the aerodynamic drag, and the method selected often dictates (constrains) many of the other operational aspects of the drop tower. The options available to reduce or eliminate the aerodynamic drag include a) dropping in a vacuum, b) dropping inside a drag shield, c) guided motion where the falling object's acceleration is matched to Earth's gravity, and certain other d) enhanced technologies (free flyers, catapults, etc.). These various methods and their specific attributes are discussed below.

6.3 Vacuum (or Drop) Tubes

Some of the earliest vacuum (or drop) tubes were used to produce commercial outcomes in the late eighteenth century for the production of high-quality spherical lead shot. Within a drop tube, the effects of aerodynamic drag are typically removed by evacuating the entire tube. This evacuation eliminates the possibility of any air drag developing on the test sample, thus allowing it to continually accelerate at 1 g during the drop. While the sample or experimental platform is in free fall, it is weightless and the effects of gravity on the phenomena being investigated can then be determined. Some drop tubes are relatively small (<1 m) in diameter and drop the test sample itself (without an experimental package), while other drop tubes are relatively big (several meters in diameter) and able to drop very large, complex experimental platforms which, however, may require a long pretest time to remove the atmosphere (air) present so that the aerodynamic drag is eliminated.

The duration of the free fall provided within a specific drop tube is directly related to the initial height of the drop through Newton's law as follows:

$$x = x_0 + v_0 t + \frac{1}{2} a t^2 \tag{6.1}$$

where t is the time of free fall, x is the distance travelled, a is the acceleration (in our case, a is the Earth's acceleration of g), and x_0 and v_0 are the initial height and velocity (v_0 is typically zero). This relation shows why all ground-based drop facilities (including vacuum tubes) are only designed to provide short durations (2–10 s) of reduced gravity. For free fall with zero initial velocity, the free fall duration is given by

$$t = \sqrt{\frac{2h}{g}} \tag{6.2}$$

where $h = x - x_0$ is the total distance dropped. This shows that doubling the height increases the free fall time by a factor of $\sqrt{2}$ only. In a catapult mode, where the experiment is initially launched upward, the free fall time doubles.

6.4 Experiment Inside Capsule (Drag Shield)

Another way to eliminate or reduce the aerodynamic drag on an object, and the operational principle some ground-based facilities are based upon, is the utilization of a "drag shield." In this configuration, shown schematically in Figure 6.1, the experimental platform is placed in a capsule or drag shield. During a test, the capsule experiences the aerodynamic drag developed as it drops through the air; however, the experiment that is inside

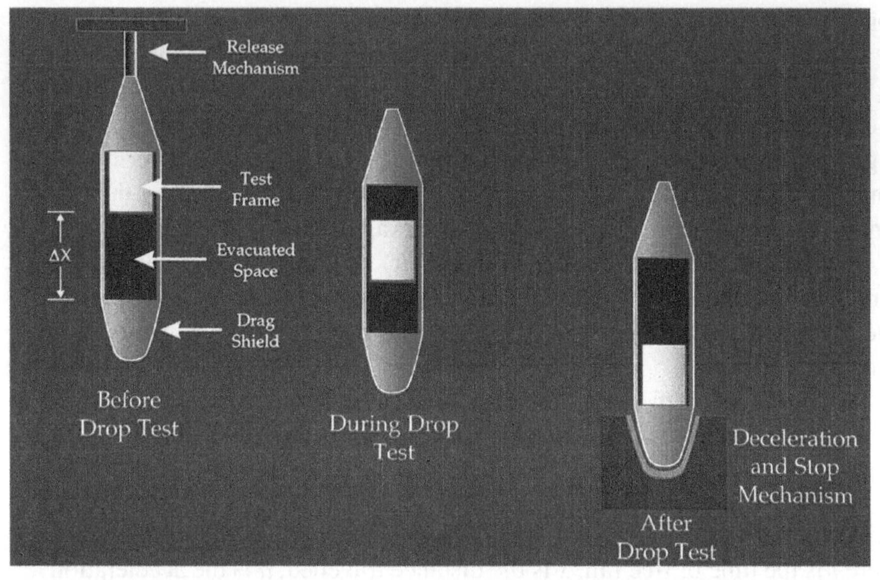

Figure 6.1 Use of drag shield to eliminate aerodynamic drag on experiment to produce reduced gravity conditions.

the capsule does not experience significant air resistance since it is only falling a short distance relative to the capsule. In this way, the aerodynamic drag on the capsule does not significantly affect the experimental platform, essentially in free fall, inside the capsule over the course of the test.

It is now interesting to know how large the initial spacing ΔX between the capsule and the bottom of the experiment platform (as shown in Figure 6.1) has to be for a free fall time of several seconds. For that, we calculate the difference of the perfect free fall given by Equation (6.1) and a fall within air where the drag shield experiences Stokes and Newton air friction. The equation of motion for this is as follows:

$$m\dot{v} = mg - \beta v - kv^2, \tag{6.3}$$

where β is the Stokes air friction coefficient and k is the Newton air friction coefficient. This equation can be solved exactly and gives

$$v(t) = -\frac{\beta}{2k} + \sqrt{\frac{\bar{g}m}{k}}\tanh\left(\sqrt{\frac{\bar{g}k}{m}}t + \tanh^{-1}\frac{\beta}{2\sqrt{k\bar{g}m}}\right) \tag{6.4}$$

where $\bar{g} := g\left(1 + \frac{\beta^2}{4kmg}\right)$. This is the velocity of the drag shield as function of time. The terminal velocity is $v_\infty := \lim_{t\to\infty} v(t) = -\frac{\beta}{2k} + \sqrt{\frac{\bar{g}m}{k}}$ from which we recover the well-known cases for $\beta \to 0$ and $k \to 0$. For the position of the drag shield, we then obtain

$$x(t) = -\frac{\beta}{2k}t + \frac{m}{k}\ln\left(\sqrt{1 - \frac{\beta^2}{4k\bar{g}m}}\cosh\left(\frac{\bar{g}k}{m}t + \tanh^{-1}\frac{\beta}{2\sqrt{k\bar{g}m}}\right)\right) \tag{6.5}$$

The spacing between the drag shield and the free flyer (test frame) then is $\Delta x(t) = \frac{1}{2}gt^2 - x(t)$. An expansion for short times gives

$$\Delta x(t) = \frac{\beta g}{6m}t^3 + \frac{g^2k}{12m}\left(1 - \frac{\beta^2}{2gkm}\right)t^4 + o(t^5). \tag{6.6}$$

The maximum possible time of flight is given by $\Delta x(t_{\max}) = \Delta X$. The result depends on the coefficients β and k which also depend on the air viscosity and the geometry of the drag shield. According to this result, most ground-based drop facilities are only designed to provide short durations of reduced gravity since the spacing required for long drops becomes much larger as drop time increases.

6.5 Drop Tower Systems

As drop towers permit entire experimental systems/platforms to be dropped, they can vary considerably in their size ranging from small drop corridors (of about 1 m diameter) to very large dimensions (of several meters in diameter). A drop tower facility, in general, will consist of (a) a drop corridor within which the experiment resides during the period of reduced gravity, (b) experimental system(s) mounted on an experimental rack unique to the phenomena being investigated and the facility the test is to be conducted in, (c) some method to produce reduced gravity conditions for the experimental platform, (d) a lifting mechanism for the experimental platform, (e) a holding and release mechanism for the experimental platform, (f) a deceleration device to stop the experimental package at the conclusion of a test, and (g) space to prepare experiments and interact with/provide access to the drop corridor.

The drop corridor is the vertical extent within which the experiment is dropped to obtain free fall conditions. Access to the drop corridor is necessary at some location to allow the experiment to be inserted and removed, as required, and this often is associated with a laboratory area for researchers to prepare their experiments. Typically, access to the experiment within the drop corridor is provided at the start of the test. In an evacuated system such as in a drop tube, special considerations are often necessary to ensure the vacuum level in the drop corridor is not lost as the experimental system is moved into the drop corridor. The experimental system is used to contain all aspects of the test being conducted. As the experiments are typically in free fall within the drop corridor, all aspects (power, data and image acquisition, device controllers, switches, etc.) of the testing are performed remotely after the test is initiated (dropped). In the case of a system utilizing a drag shield, the experiment must be loaded into the drag shield and the resulting package readied for (and recovered from) the drop corridor as a single item. If required, a lifting mechanism is utilized for raising (pretest) or recovering (post-test) the experiment and/or experiment and drag shield. At the initiation of a test, the experiment (and drag shield, if used) must be held and released, as required, consistent with the facility requirements for test initiation. One of the most critical things during test initiation is the minimization of any unwanted vibrations (g-jitter) that may be imparted to the experiment as this unwanted acceleration can, depending upon the experiment, detrimentally affect the results. At the conclusion of a drop test, a deceleration device is essential to bring the experimental system (and drag shield, if used) to a controlled stop in a safe fashion.

6.5.1 Guided Motion

An alternative to the basic drop tower configuration described above is provided by a "guided motion" drop system. In a guided motion drop system, there is no need for a drag shield as the experiment is contained in a capsule that is then propelled downward at an acceleration of 1 g. The guided motion of the capsule is typically obtained by rail guides and/or levitated drive devices (similar to high-speed trains). By matching the Earth's gravitational acceleration, the capsule is essentially in free fall and in a 0 g environment. The matching of the capsules' acceleration to Earth's gravity vector is normally accomplished by having the capsule move along a predefined velocity profile (easily derived through the use of Equation 6.1). Since the velocity (and, therefore, the associated distance travelled) increases significantly with time, these facilities are typically limited to between 5 and 10 s of test time.

6.6 Enhanced Technologies

Several technologies are used to further increase either the duration (time) or quality (level of reduced gravity) of the test environment within the systems described above and are presented here.

6.6.1 Free Flyer System

Precise experiments often require an enhanced environment which sometimes exceeds the given quality level of reduced gravity in an evacuated drop tube or in a drag shield. In general, the quality of reduced gravity depends on the residual air pressure present. These drop systems will typically have several hundreds up to a few thousands of cubic meters of air to evacuate. From an operational point of view, since the vacuum pumps utilized have limited pump speeds, there will often remain a pressure of a few mbar in the drop corridor. For the drop tower Bremen, there is about 0.1 mbar present that allows excellent microgravity conditions of about 10^{-6} g to be produced. In order to further reduce the gravity level for special precision experiments, a free flyer system is applied [1]. The concept of the free flyer is based on the utilization of a drag-shielded experiment as previously described which is dropped inside an evacuated drop tube. In this application, the free flyer system effectively minimizes residual disturbances being caused by the low-level aerodynamic drag still present. At the drop tower Bremen, experiments utilizing the free flyer system can achieve reduced gravity conditions of about 10^{-7} g.

6.6.2 Catapult System

An increase in the time of reduced gravity available in a drop tower can obviously be obtained by simply building a higher drop tower or longer drop shaft. Due to the fact that, however, the time of reduced gravity available is a function of the distance dropped to the ½ power (Equation 6.2), construction of drop towers or drop shafts does not seem to be very economical after reaching a certain drop distance. A less utilized approach that essentially doubles the test time for a given drop distance is to catapult the experiment upward from the bottom of the drop corridor to make both the way up and the way down available (this is an especially attractive option in an existing drop facility since no further height is required). Although the experiment is initially accelerated after its detachment from the acceleration unit (i.e., the catapult), it performs a free fall in a vertical parabola (comparable to parabolic flights). Either in an evacuated drop tube or in a drag shield configuration, a catapulted experiment becomes a free-falling body and is weightless.

Accelerating an experiment capsule carrying hundreds of kilograms of payload requires a powerful drive coupled with precise control and handling of the capsule movement which is essential during the acceleration phase and to ensure exact capsule alignment during the flight phase within the confines of the drop corridor. At the moment, the drop tower Bremen is the only ground-based facility with a catapult system [2]. This world unique catapult system is installed in a chamber below the basement and utilizes a combined hydraulic–pneumatic drive to accelerate an experimental capsule (with a total mass of over 400 kg) to the top of the evacuated drop tube on a vertical parabola. The pressure difference between the vacuum level present inside the drop tube and the compressed air below the catapult piston (stored in pressurized air tanks) is utilized as the driving force of the catapult system. The acceleration level of the catapult (with up to 30 g's achieved within about 300 ms) is adjustable by a servo-hydraulic braking system that provides smooth control of both the overall piston velocity and the final transition of the experiment to microgravity at release. Typically, an experimental capsule will leave the catapult system with a lift-off speed of the about 48 m/s at 10^{-6} g with minimal vibration from the catapult.

6.6.3 Next-Generation Drop Towers

Drop facilities represent an important economic alternative with straightforward and permanent access to weightlessness on Earth in comparison with the other possible flight opportunities. In recent years, numerous demands of higher repetition rates for experiments under weightlessness have been observed at the available ground-based facilities. The reasons are, on the one hand, that current experiments become more and more complex and they require, therefore, higher repetition rates to generate the specific number of experiment parameters for their success. On the other hand, experiments need reliable statistics in their experimental results which have been achieved so far. Both scientific aspects and additionally the continuous technological progress in the present experiment developments (i.e., fully autonomous computer operations) lead to a magnitude of demand that exceeds the given capabilities afforded by existing drop facilities. The limiting factors of achieving higher repetition rates and thus laboratory-closed conditions at the present drop facilities include the time it takes to evacuate the drop tube and experiment capsule recovery and preparation for the next drop. At the drop tower Bremen, a second drop facility is currently planned which offers repetition rates of over 100 experiments per day in a semi-continuous laboratory operation. This next-generation drop tower system, the so-called GraviTower Bremen, combines the technological benefits of the catapult system of the drop tower Bremen with a guided electromagnetic linear drive. The new facility concept is based on the operation of a commercial elevator in which the distance between passenger (i.e., the experimental payload a free flyer) and cabin walls (drag shield) is actively regulated. After each initial acceleration, and subsequent detachment, the experimental payload becomes a free-flying body and is weightless. In order to smoothly accelerate and decelerate an experimental payload of several hundred kilograms on a vertical parabola in such a guided catapult system, it is essential to work with a both powerful and precisely controllable drive. An electromagnetic linear drive commonly utilized in roller coasters represents a commercially available and well-tested solution for the semi-continuous experiment operation. As currently planned, the GraviTower Bremen will have a height of about 70 m and will provide fall durations of 6 s at microgravity conditions. The different free fall durations will depend on the selectable acceleration/deceleration drive mode with a range available between 1.5 and 4.0 g. Only very low initial experiment disturbances are expected during the acceleration phase. The accuracy in power and control of the applied electromagnetic linear drive allows also a novel operation mode with a fixed "free flyer." In this case, experiments under partial gravity (at gravity levels from 0.1 to 0.4 g with an expected accuracy of 10^{-2} g) can be conducted. The duration of partial gravity available at the GraviTower Bremen ranges from 5.5 to 7.8 s and depends on both the partial g-level and drive mode selected. Finally, the dimensions of the GraviTower's free flyer are planned to handle much larger volume and mass than in other existing facilities.

6.6.3.1 Ground-based facility's typical operational parameters

In the characterization of a ground-based facility, four factors are most significant: test time, magnitude of reduced gravity (quality), size of experiment that can be accommodated, and cost. Other parameters are sometimes considered such as facility location, deceleration experienced at test conclusion, and technical assistance available. These characteristics are function of the facility being considered and the methodology it uses to produce the reduced gravity. Examples of these characteristics are provided and contrasted in Table 6.1.

Table 6.1 Characteristics of a large and a small ground-based reduced gravity

Facility	Location	Test Time (s)	g-Level (g's)	Approximate Size of Experiment	Approximate Cost	Deceleration Level at Test Conclusion
ZARM	Bremen, Germany	4.7 or 9.3 (with catapult)	$10^{-6}/10^{-7}$ (with free flyer)	Cyl.: up to 0.8 m dia., up to 1.7 m long, about 300 kg (different exp. capsules)	Varies	Up to 50 g (typical 40 g)
NASA	Cleveland, USA	5.2	10^{-6}	Cyl.: up to 1 m dia., up to 1.6 m high, up to 455 kg (incl. drop vehicle)	Varies	up to 65 g (mean 35 g)
NML	Beijing, China	3.6	$10^{-3}/10^{-5}$ (with free flyer)	Varies	Varies	15 g
QUT	Brisbane, Australia	2.0	10^{-4}–10^{-6}	Cyl.: 0.9 m dia × 1.5 m long	€400–€600	15–20 g's for ~0.25 s

6.7 Research in Ground-Based Reduced Gravity Facilities

There are many scientific discipline areas that currently use ground-based facilities to produce reduced gravity test conditions to study relevant phenomena. A short description of some of the work being conducted is presented.

6.7.1 Cold Atoms

Quantum objects show interference: If a beam of quantum objects is split coherently and the two beams are moving along different paths, then after recombination of both beams, interference will occur. The interference pattern is influenced by external influences like acceleration or rotation of the whole interferometer or by the influence of other external forces. The important issue is that the effect on the interference scales with the square of the time the atoms move within the interferometer. While on ground, the atoms will fall down the table within a tenth of a second, we can increase the time if the interferometer will fall together with the atoms. In a free fall environment, interferometers become much more sensitive. Since after a few seconds, the coherence of the two beams disappears due to imperfect isolation against disturbances, drop towers are currently ideal facilities to perform such experiments [3, 4]. A future goal of this ground-based work is to bring cold atoms to space, that is, on the ISS or on dedicated spacecraft. This work supports fundamental physics experiments such as quantum tests of the equivalence principle, tests of the linearity of quantum mechanics, or practical applications such as geodesy and Earth sciences.

6.7.2 Combustion

Combustion experiments in many diverse areas are being studied. Some studies are aiming at a thorough investigation of the self-ignition of droplets of different types of fuel, including modern biofuel of the second generation [5]. These investigations form the foundations of the understanding of the physical processes underlying spray ignition needed for efficient and ecologically compatible combustion. The unique heterogeneous (liquid phase) burning of bulk metallic materials in oxygen-enriched atmospheres is also being studied with specific applications to fire safety and fundamental combustion science [6, 7]. For these experiments, weightlessness is of big advantage because (i) convection then does not occur, (ii) the system is simplified due to the spherical symmetry of the liquid-phase droplets, and (iii) one can study larger droplets which often permit a more detailed observation of the processes being studied. One main result from the obtained data is input to assist the development of a computer simulation for the detailed study of the complete spray ignition process. This has a big impact on the construction of energetically efficient and ecologically compatible engines.

6.7.3 Fluid Mechanics/Dynamics

It is obvious that fluids behave differently according to whether gravity is present or not. One effect is related to the surface tension which makes it possible, in the absence of gravity, for large liquid balls to be formed. Another effect is related to capillarity: For very weak gravity, the effect of adhesion of water at surfaces becomes dominant with the effect that a fluid will move along surfaces. A particular topic being investigated is the behavior of free surfaces [8]. Such an effect can be used for propellant management devices or life support systems in satellites.

Special topics under investigation include multiphase systems and cryogenic fluids including the capillary channel flow (CCF) project, a two-phase system which has been utilized under microgravity conditions on the ISS and which required a significant number of preparatory investigations conducted in drop tower experiments [9, 10]. Cryogenic propellants will be used in the Ariane V upper stage; therefore, it is very important to explore the behavior of these fluids in microgravity to elucidate the underlying physical principles. These principles will be used to develop a computer simulation tool for the construction of thrusters.

Another active area of research is related to the investigation of boiling and heat transfer under microgravity conditions [11]. Novel configurations of bubble suspensions and heat transfer mediums (some incorporating nanoparticle suspensions) are created to study coalescence, phase change dynamics, and heat exchange in general in a turbulent medium under controlled conditions. This combines the physics of two-phase flows and thermal control which has importance for technology applications in both space and terrestrial applications.

6.7.4 Astrophysics

A main task of the experiments supporting astrophysics work is to determine parameters needed to estimate the characteristic time for the formation of planets, moons, and other solar system objects [12]. In particular, studies are conducted about the conditions under which elastic and inelastic scattering of dust particles, granular particles, small ice particles, or others occurs to form larger constituents (agglomeration). This phenomenon depends on the size and mass of the particles, their temperature, etc. Other external influences, such as light or thermal radiation, are also being investigated. Microgravity experiments revealed that the surface of Mars is efficient in cycling gas through layers, at least centimeters above and below the soil, with a turnover time of only seconds to minutes [13]. Clearly, such experiments must be conducted under microgravity conditions to properly simulate the conditions in space—planets and other objects that form under weightlessness condition. Besides this type of experiment's relevance to planetary sciences, this work may also have impact on material sciences, in particular, on the physics of granular materials.

6.7.5 Material Sciences

Particular problems in the area of material sciences such as the physics of granular gases, synthesis of nanomaterials, or the transport of fluids in porous materials are also being studied in microgravity environments. One aspect is to discover fundamental properties and synthesis pathways within these systems which are relevant for producing novel materials not able to be synthesized in normal gravity and supporting theoretical/statistical descriptions of these systems (and/or validating corresponding numerical simulations) [14, 15]. Another application of this work is for the transport of granular gases and of the transport of fluids through granular or porous media which has industrial applications including satellite technology. For satellite technologies, cryogenic fluids must be considered and properly characterized. This work is conducted in microgravity conditions in order to explore the fundamental physical principles without the disturbing gravitational force. The impact and relevance for industrial applications is obvious.

6.7.6 Biology

The experimental study of biological systems under microgravity or reduced gravity is in some cases of general interest but mainly for applications in space, on Moon, or on Mars to better understand how do organisms behave and how the nervous system reacts. In drop towers, only phenomena occurring on a short timescale can be investigated and this tends to limit biological applications. This includes the general behavior of microorganisms or the influence of microgravity on the orientation capabilities of complex organisms [16]. This work supports a better understanding of the general functioning of the nervous system but also is very important for estimating the reaction of life during long space travel or for its behavior under variable gravity conditions (Moon, Mars, etc.).

6.7.7 Technology Tests

Drop towers are also used for technology tests to operate and validate various systems prior to deployment in operational environments. Examples are tests of the behavior of heat pipes, the performance of accelerometers, or the functioning of release mechanisms. The full performance of high-precision accelerometers can only be explored under weightlessness conditions. At the drop tower Bremen, testing is conducted in support of the differential accelerometers for the French mission MICROSCOPE, aimed at testing the universality of free fall (also called weak equivalence principle), with an accuracy of 10^{-15}, which is two orders of magnitude better than what is possible on Earth. The accelerometers, built by ONERA in Paris, have a performance of 10^{-14} m/s^2/sqrt(Hz). For these accelerometers, the catapult of the Bremen drop tower with a free fall time of almost 10 s is used [17]. Also, the accelerometers to be used in the GRACE Follow-On geodesy mission will be tested in the Bremen drop tower.

Another technology validation program is the asteroid lander MINERVA (MIcro/Nano Experimental Robot Vehicle for Asteroid) of the Japanese Hayabusa 2 mission tested in the Bremen drop tower. The actual mechanism for taking soil samples that are planned to be brought to Earth has been tested and validated for functionality. In addition, the functioning of the release mechanism of mobile lander MASCOT (Mobile Asteroid Surface Scout) has been proven in the Bremen drop tower as cooperation projects between the German Aerospace Center DLR and the Japanese space agency JAXA. A last example of technology validation would include the tests of the heat pipes designed to cool the eROSITA camera system at a temperature of –95 °C. The validation testing demonstrated that the heat pipe system that was based on capillary forces is working very well and should perform well once deployed.

References

[1] Selig, H., H. Dittus and C. Lämmerzahl. “Drop Tower Microgravity Improvement Towards the Nano-g Level for the MICROSCOPE Payload Tests”. *Microgravity Science Technology* 22 (2010): 539.

[2] von Kampen, P., U. Kaczmarczik and H.J. Rath. "The New Drop Tower Catapult System". *Acta Astronautica* 59 (2006): 278.

[3] van Zoest, T. et al. "Bose–Einstein Condensation in Microgravity". *Science* 328 (2010): 1540.

[4] Müntinga, H. et al. "Interferometry with Bose Einstein Condensates in Microgravity". Phys. Rev. Lett. 110 (2013): 093602.

[5] Burkert, A., W. Paa, M. Reimert, K. Klinkov and C. Eigenbrod. "Formaldehyde LIF Detection with Background Subtraction Around Single Igniting GTL Diesel Droplets". *Fuel* 111, no. 8 (2013).

[6] Lynn, D.B., O. Plagens, M. Castillo, T. Paulos and T.A. Steinberg. "The Increased Flammability of Metallic Materials Burning in Reduced Gravity", In *Proceedings of the 4th International Association for the Advancement of Space Safety Conference: Making Safety Matter*, ESA-SP Volume 680, (2010), pp. 1–7.

[7] Ward, N.R. and T.A. Steinberg. "Iron Burning in Pressurised Oxygen Under Microgravity Conditions." *Microgravity Science and Technology* 21, no. 1–2 (2009): 41–46.

[8] Diana, A., M. Castillo, D. Brutin and T. Steinberg. "Sessile Drop Wettability in Normal and Reduced Gravity". *Microgravity Science and Technology* 24, no. 3 (2012): 195–202.

[9] Canfield, P.J., P.M. Bronowicki, Y. Chen, L. Kiewidt, A. Grah, J. Klatte, R. Jenson, W. Blackmore, M.M. Weislogel, and M.E. Dreyer. "The Capillary Channel Flow Experiments on the International Space Station: Experiment Set-Up and First Results". *Experiments in Fluids* 54, no. 1519 (2013): 1.

[10] Conrath, M., P.J. Canfield, P.M. Bronowicki, M.E. Dreyer, M.M. Weislogel and A. Grah. "Capillary Channel Flow Experiments Aboard the International Space Station". *Physcal Review E* 88 (2013): 063009.

[11] Diana, A., M. Castillo, T.A. Steinberg and D. Brutin. "Asymmetric Interface Temperature During Vapor Bubble Growth". *Applied Physics Letters* 103, no. 3 (2013).

[12] Blum, J. "Astrophysical Microgravity Experiments with Dust Particles". *Microgravity Science Technology* 22 (2010): 51.

[13] de Beule, C., G. Wurm, T. Kelling, M. Kuepper, T. Jankowski and J. Teiser. "The Martian Soil as a Planetary Gas Pump". *Nature Physics* 10 (2014): 17.

[14] Bánhidi V., and T.J. Szabo. "Detailed Numerical Simulation of Short-Term Microgravity Experiments to Determine Heat Conductivity of Melts". *Materials Science Forum* 649 (2010): 23.

[15] Hales, M.C., T.A. Steinberg and W.N. Martens. "Synthesis and Characterization of Titanium Sol–Gels in Varied Gravity". *Journal of Non-Crystalline Solids* 396–397 (2014): 13–19.
[16] Anken, R. and R. Hilbig. "Swimming Behaviour of the Upside-Down Swimming Catfish (*Synodontis nigriventris*) at High-Quality Microgravity—A Drop-Tower Experiment". *Advances in Space Research* 44 (2009): 217.
[17] Seilg, H. and C. Lämmerzahl. "Sensor Calibration for the MICROSCOPE Satellite Mission". *Advanced Astronomical Science* 148 (2013).

7

Parabolic Flights

Vladimir Pletser[1] and Yasuhiro Kumei[2]

[1]ESA-ESTEC, Noordwijk, The Netherlands
[2]Tokyo Medical and Dental University, Tokyo, Japan

7.1 Introduction

Aircraft parabolic flights are useful for performing short-duration scientific and technological experiments in reduced gravity. Their principal value is in the verification tests that can be conducted prior to space experiments in order to improve their quality and success rate, and after a space mission to confirm (or invalidate) results obtained from space experiments. Parabolic flight experiments might also be sufficient as stand-alone, addressing a specific issue.

The levels of reduced gravity that can be attained during parabolic flights vary between 0 and 1 g (where g is the acceleration created by gravity at Earth's surface, on average 9.81 m/s^2). Near weightlessness at micro-g (μg) levels is commonly achieved during ballistic parabolic flights for microgravity research since more than 30 years. Recent years have seen more and more emphasis on partial-g flight profiles allowing to obtain gravity levels similar to those on the Moon (0.16 g) and on Mars (0.38 g) to prepare for future space exploration.

This section introduces the objectives of parabolic flights for research and the flight profiles to achieve μg and partial-g environments. The various airplanes used throughout the world to create these reduced gravity environments are introduced. These airplanes can be grouped in three categories: (1) large airplanes allowing to embark several tens of passengers and several large experiments; (2) medium-sized airplanes usually used for single experiment with several operators and/or subjects; and (3) small airplanes and jets embarking single passengers and small experiments.

7.2 Objectives of Parabolic Flights

A large aircraft in parabolic flight provides investigators with a laboratory for scientific experimentation where the g-levels are changed repetitively, giving successive periods of either 0.38 g for up to 32 s, or 0.16 g for up to 25 s or μg for 20 s, preceded and followed by periods of 20 s at approximately 1.8 g-level.

Parabolic flight objectives pursued by scientists are usually multifold: (1) to perform short experiments for which the reduced gravity is low enough for qualitative experiments of the "look-and-see" type, using laboratory-type equipment to observe and record phenomena, and quantitative experiments to measure phenomena in reduced gravity, yielding direct quantitative exploitable results; (2) to allow experimenters to perform by themselves their own experiments in reduced gravity with the possibility of direct interventions on the experiment in progress during the low g periods and direct interaction by changing experiment parameters between the reduced gravity periods; and (3) to study transient phenomena occurring during changeovers from high-to-low and low-to-high g-phases. Additionally, for space mission experiments, preliminary results can be obtained prior to a space mission and experiments with conflicting results can be repeated shortly after a space mission, helping in data interpretation. For space human physiology experiments on astronauts, a broader data baseline can be obtained in μg prior to or after a space mission by conducting parts of the space experiments on a group of subjects other than astronauts.

From a technical point of view, in preparing experiment hardware for manned spaceflight or robotic missions, the following objectives can also be achieved: (1) test of equipment hardware in reduced gravity; (2) assessment of the safety aspects of an instrument operation in reduced gravity; and (3) training of science astronauts to experiment procedures and instrument operation.

Furthermore, aircraft parabolic flights are the only suborbital carrier to provide the opportunity to carry out medical and physiological experiments on human subjects in μg or partial gravity at Moon and Mars g-levels, to prepare for extraterrestrial planetary exploration.

Finally, other g-levels achieved during flights can be used by investigators: Pull-up and pullout maneuvers yield periods of hypergravity (from 1.8 g to several g's); spiral turn maneuvers provide for longer periods of other levels of high g's.

7.3 Parabolic Flight Maneuvers

For μg parabolic flights, the μg environment is created in a large aircraft flying the following maneuvers (see Figure 7.1, example given for the Airbus A300 ZERO-G):

- from steady horizontal flight, the aircraft climbs at approximately 45° (pull-up) (see Figure 7.2) for about 20 s with accelerations between 1.8 and 2 g;
- all aircraft engine thrust is then strongly reduced for about 20 to 25 s, compensating the effect of air drag (parabolic free fall);
- the aircraft dives at approximately 45° (pullout), accelerating at about 1.8 to 2 g for approximately 20 s, to come back to a steady horizontal flight.

Alternatively, for partial g parabolas, the engine thrust is reduced sufficiently to a point where the remaining vertical acceleration in the cabin is approximately 0.16g for approximately 25 s or 0.38 g for approximately 32 s with angles at injection of 42° and 38°, respectively, for Moon and Mars parabolas. These maneuvers can be flown consecutively (in a roller coaster manner) or separated by intervals of several minutes to allow investigators to prepare for their experiments.

For μg parabolas, the residual accelerations sensed by experimental setups attached to the aircraft floor structure are typically in the order of 10^{-2} g, while for a setup left free floating in the cabin, the levels can be improved to typically 10^{-3} g for 5 to 10 s.

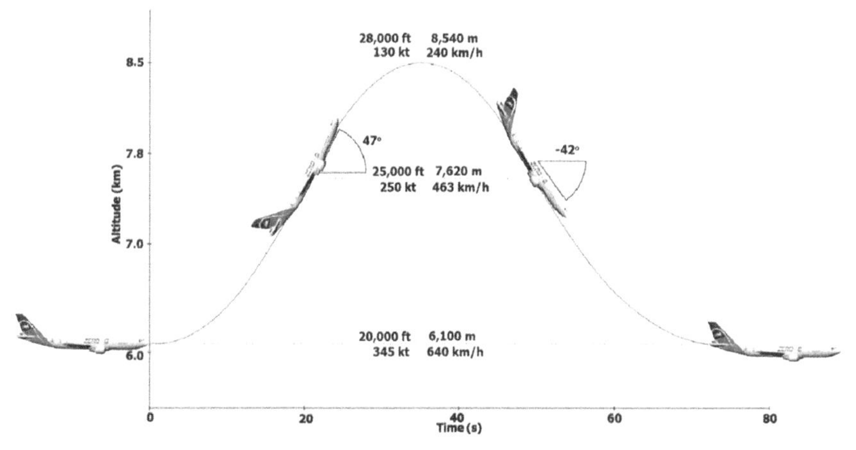

Figure 7.1 Airbus A300's parabolic flight maneuver (Credit: Novespace).

Figure 7.2 The Airbus A300 in pull-up (Photo: Novespace - Eric Magnan/Airborne Films).

7.4 Large Airplanes Used for Parabolic Flights

Large airplanes used for parabolic flights are defined as those aircraft used for flying several (typically ten or more) experiments and embarking several tens of passengers, either for research purposes or for discovery reduced gravity flights for paying passengers.

7.4.1 Europe: CNES' Caravelle and CNES-ESA's Airbus A300 ZERO-G

From 1988 till 1995, the European Space Agency (ESA) and the "Centre National d'Etudes Spatiales" (CNES, French Space Agency) have used a Caravelle aircraft for several tens of campaigns [1].

From 1997 till 2014, the Airbus A300 ZERO-G was used in Europe by ESA, CNES, and the "Deutsches Zentrum für Luft- und Raumfahrt e.V.," DLR, the German Aerospace Center, and industrial users for μg and partial-g flights. The Airbus A300 was the largest airplane in the world used for this type of experimental research flight.

Since 2015, an Airbus A310 is used to replace the A300 ZERO-G. The French company Novespace, a subsidiary of CNES, based in Bordeaux, France, is in charge of the organization of Airbus A300 and A310 flights.

Prior to a campaign with an Airbus ZERO-G, space agencies and Novespace provide support in the experiment equipment design and in all related safety aspects. All experiments are reviewed by experts several months before a campaign from the structural, mechanical, electrical, safety, and operational points of view. Technical visits are made to the experimenters' institutions to review equipment. A safety review is held one month before the campaign. A safety visit is made in the aircraft prior to the first flight to verify that all embarked equipment complies with the safety standards. The campaign in itself takes place over two weeks. The first week is devoted to the experiment preparation and loading in the aircraft. During the second week, on the Monday, a safety visit takes place to assess that all safety recommendations have been implemented and a flight briefing is organized in the afternoon to present the flight maneuvers, the emergency procedures and medical recommendations, and the experiments on board. The three flights of 30 parabolas each take place on the mornings of the Tuesday, Wednesday, and Thursday followed each time by a debriefing during which the needs and requests of investigators are reviewed and discussed. Due to bad weather or technical problems, a flight can be postponed from the morning to the afternoon or to the next day. Downloading of all experiments takes place on the afternoon after the last flight.

A typical flight duration with the Airbus ZERO-G is about two and half hours, allowing for 30 parabolas to be flown per flight, in sets of five with two-minute intervals between parabolas and with four to six minutes between sets of parabolas. Parabolas are flown in dedicated air zones over the Gulf of Biscay or the Mediterranean Sea.

Since 1984, a total of 132 European campaigns were performed (61 ESA, 48 CNES, 23 DLR) for more than 1500 selected experiments in human physiology and medicine, biology, physics, astrophysics, technological tests, and launcher technology proposed by researchers and students [2–5] (see Figure 7.3). In addition to μg campaigns, two joint European partial-g campaigns were organized by ESA, CNES, and DLR in 2011 and 2012 for experiments at μg and lunar and Martian gravity levels [6].

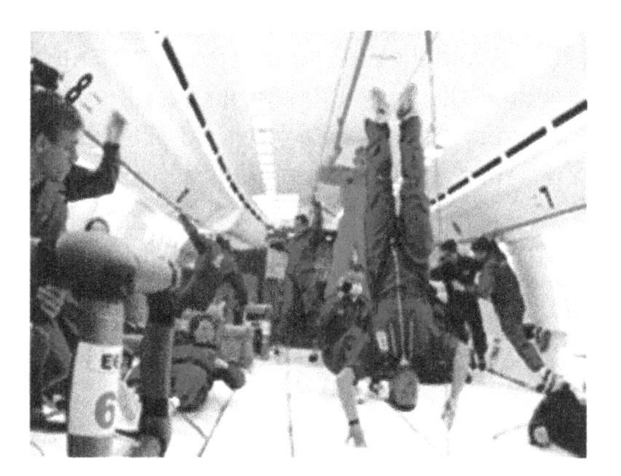

Figure 7.3 Experimenters during μg parabolic flights on the airbus A300 (Photograph ESA).

Since 2013, the Airbus A300 and A310 ZERO-G are used for discovery flights open to the public.

7.4.2 USA: NASA's KC-135, DC-9 and Zero-G Corporation

A review of the aircraft used by NASA for parabolic flights since the early 1950 is given in [6].

NASA has operated several KC-135 aircraft (modified Boeing 707 jet aircraft) for reduced gravity research and astronaut training. The KC-135A/930, named "Weightless Wonder IV" (and nicknamed "Vomit Comet"), was the longest operated aircraft (1973–1995) [7]. ESA used the KC-135A aircraft for its first six μg research campaigns (1984–1988) [8], and DLR conducted nine campaigns (1987–1992) in preparation for the German Spacelab D2 mission. The replacement aircraft, the KC-135A/931, was used until its retirement in 2004 [9].

NASA operated also a DC-9 to allow researchers to perform their experiments in a reduced gravity environment. A flight lasts typically 2.5 h, with 40–60 parabolas, during which μg and partial gravity levels (0.16, 0.38 g) and sustained hyper-g (1.6 g) can be achieved according to researcher requirements [10].

The private company Zero Gravity Corporation operated a modified Boeing 727-200, named "G-FORCE ONE" in the USA from 2004 until 2014 for discovery flights open to the public. NASA had a microgravity service contract with Zero Gravity Corporation from 2008 until 2014 to fly NASA's sponsored experiments [11]. The flight maneuver is quite similar to the one described in earlier sections. From an horizontal flight at an altitude of approx. 8000 m, the pull-up up to 1.8 g lasts about 10 to 17 s up to an altitude of approx. 11300 m, and the aircraft is injected into the parabola for a duration of approx. 20 s until the pullout takes place. A flight includes 12–15 parabolas for public discovery flights or 25–40 parabolas for research flights. Moon-g and Martian-g flight maneuvers can also be performed [12].

7.4.3 Russia: Ilyushin IL-76 MDK

The Ilyushin IL-76 MDK (MDK stands for "latest modifications" in Russian) is a four-jet-engine cargo aircraft of the last modified version [13] and operated by the Russian Yu. Gagarin Cosmonaut Training Centre (CTC) from Star City near Moscow and used for parabolic flights for astronaut training, space equipment tests, and paying passenger flights. The main aircraft features are the double-floor cockpit and a large cabin, separated into two parts: The front includes several work stations for experimenters conducting tests on large equipment, while the aft part is an empty cabin space with attachment points on the cabin floor. ESA and DLR conducted parabolic flight campaigns with this aircraft in 1992 and 1994 [14]. The Ilyushin IL-76 MDK is marketed since the nineties by several private operators for discovery flights open to the public.

7.5 Medium-Sized Airplanes Used for Parabolic Flights

Medium-sized airplanes are defined as those aircraft used for flying single experiments with several operators and/or subjects.

7.5.1 Europe: TU Delft-NLR Cessna Citation II

The Cessna Citation II is a twinjet aircraft, a research aircraft owned and operated jointly by the Technology University of Delft and the Dutch National Aerospace Laboratory in the Netherlands. It has been extensively modified to serve as a versatile airborne research platform. The flight envelope with a maximum altitude of more than 13 km allows a wide range of operations to be performed, including flying at quite low speeds. It can accommodate a maximum of eight observers in addition to the two-pilot cockpit crew [15]. Among other research flights, it is used for aerospace student practical training, for parabolic flights for single experiments, and for discovery flights open to the public.

7.5.2 Canada: CSA Falcon 20

The Falcon 20 is a twin-engine business jet, capable of relatively high-speed and altitude operations with a small complement of instrumentation and research crew. It had been modified for μg experiments requiring parabolic flight trajectories [16]. The aircraft, owned and operated by the National Research Council's Institute for Aerospace Research (NRC/IAR), was used until 2014 by the Canadian Space Agency (CSA) for Canadian investigators.

7.5.3 Japan: MU-300 and Gulfstream-II

In Japan, a MU-300 jet aircraft operated by Diamond Air Service since 1990 [18] provides parabolic flights with up to 20 s of weightlessness. Besides creating conventional μg conditions, the parabolic flight using this MU-300 has many advantages. Since only one or two research themes are conducted in this medium-sized aircraft, the flight maneuver can be customized flexibly so as to make the most suitable condition for the study. Experiments have been conducted on rodents and cell cultures since 2007, using not only Martian (0.38 g) and lunar (0.16 g) gravity levels but also 0.6, 0.5, 0.4, 0.3, 0.2, 0.1, and 0.05 g [19]. Other unique studies were also realized by using partial gravity conditions [20, 21]. A set of original parabolic trajectories was used to avoid interference of the "pull-up" phase-induced hypergravity into the genuine response to low gravities. After a certain time in the 1 g-level flight, the aircraft enters into the "pull-up" phase that lasts approximately 20∼40 s at 1.3 g, followed by a sudden descent that generates a target partial gravity condition lasting 15∼40 s and then pulls up again into the "recovery" phase that lasts 20 s at 1.3 g (see Figure 7.4).

Diamond Air Service operates also a jet aircraft "Gulfstream-II" since 1996 in Nagoya. The cockpit crew includes two pilots and one engineer who support the in-flight experiments. A total of 5.5-hour flight is possible

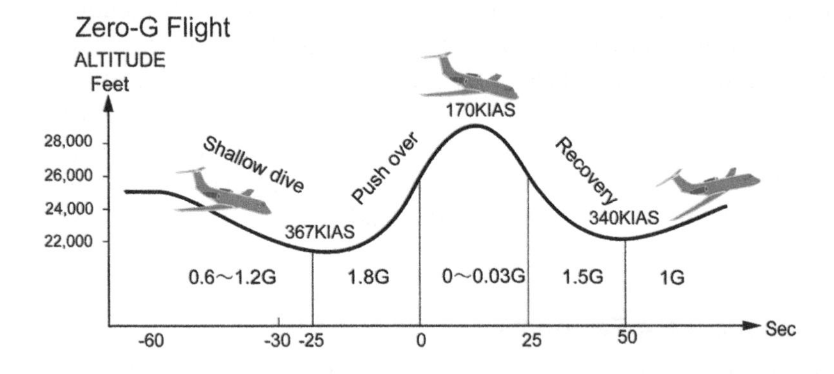

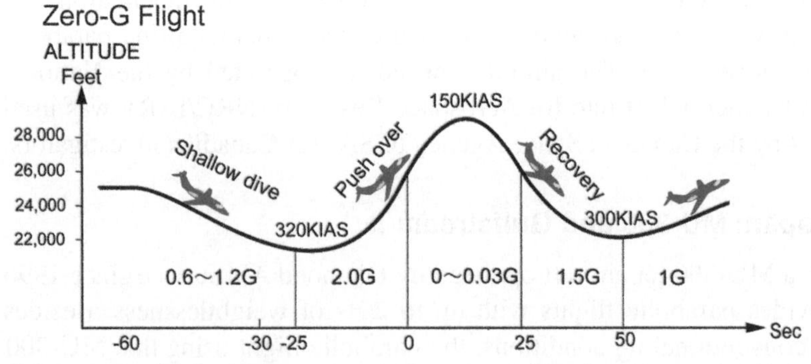

Figure 7.4 Typical μg flight trajectories of the Gulfstream-II (top) and the MU-300 (bottom) (Credit: Diamond Air Service).

with 5 or 6 researchers and up to 2.4 t of payload. To date, a total of 1480 flights with MU-300 and 350 flights with Gulfstream-II have been conducted so far, and 845 studies have been accomplished with JAXA support.

7.5.4 Other Aircraft

Ecuador: The Ecuadorian Air Force (FAE) and the Ecuadorian Civilian Space Agency (EXA) adapted jointly T-39 Sabreliner to perform μg parabolic flights [22]. In 2008, two flights took place and a fluid experiment was performed.

Austria: The company Pauls Parabelflüge organized since 2002 a series of flights in Germany, Austria, and Slovakia with several medium-sized aircraft (Casa 212, Short Skyvan SC7, Cessna Grand Caravan, Let 410) and recently with a glider (SARA-Mercury) yielding parabolas of 10 to 15 s of μg for discovery flights for paying passengers and students [22].

7.6 Small Airplanes and Jets Used for Parabolic Flights

Small airplanes and jets used for parabolic flights are defined as those aircraft used for flying single passengers and small experiments.

7.6.1 Switzerland: Swiss Air Force Jet Fighter F-5E

A jet fighter F-5E Tiger II aircraft was used in Switzerland from Emmen Air Force Base for biological research contained in a small experimental apparatus installed in the cannon ammunition box [23]. The military fighter jet aircraft Northrop F-5E "Tiger II" is a light supersonic fighter aircraft with two axial straight turbines, able to reach a maximum speed of Mach 1.64, with a maximum climb rate of 571 ft/s and a maximum altitude of 51,800 ft.

Single parabolas of up to 45 s are executed during military training flights (see Figure 7.5). After a 1 g control phase, the parabolic maneuver starts at 13,000 ft. and at Mach 0.99 airspeed, with a 22-s climb with an acceleration of 2.5 g up to an angle of 60° at an altitude of 18,000 ft., followed by a free fall ballistic trajectory lasting 45 s of 0.05 g in all axes with an apogee of 27,000 ft. at Mach 0.4 airspeed. The pullout is initiated when the aircraft reached 60° nose down with an acceleration of 2.5 g (up to 3.5 g) and lasts for approximately 13 s. Access time is 30 min before takeoff, and retrieval time is 30 min after landing.

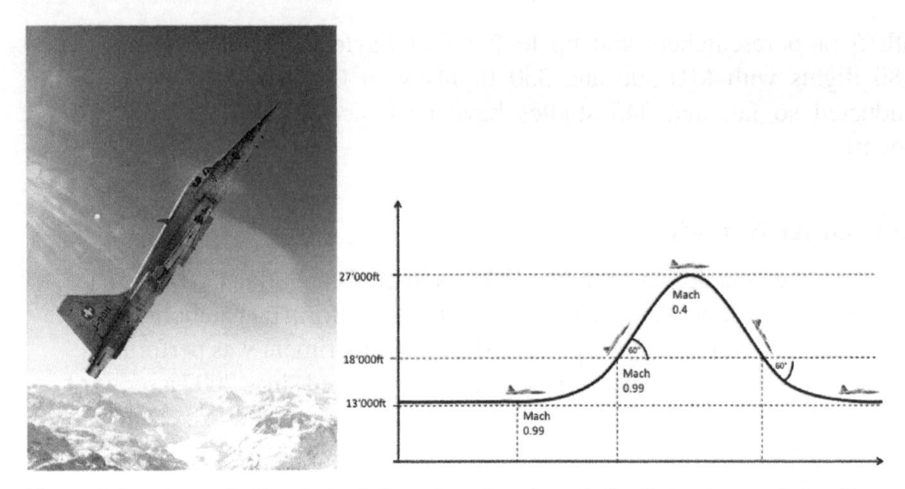

Figure 7.5 The F-5E Tiger II jet fighter aircraft and parabolic flight characteristics (from [22]).

7.6.2 Other Aircraft

Spain: In Barcelona, the Universitat Politecnica de Catalunya with the Aero Club Barcelona Sabadell proposes since 2007 parabolic flights on board a Mudry Cap10B aircraft, a two-seat training aerobatic aircraft. Up to ten parabolas of 5 to 8 s of μg each, with pull-ups and pullouts up to 3.5 g, can be performed by researchers and students with small experiments (typically 30 × 20 × 20 cm, max. 10 kg).

Belgium: In the early 1990s, a two-seat Fouga Magister from the Brustem base of the Belgian Air Force was used, as part of aerobatics training, to perform parabolic flights of duration between 15 and 24 s. It was used to conduct fluid physics and technology development experiments from the University of Brussels [24].

7.7 Conclusions

Aircraft parabolic flight maneuvers are a very useful tool to investigate gravity related phenomena, whether in complete weightlessness or at partial-g levels. As any small or large airplane could basically be used to undergo a parabolic trajectory, it is important to choose carefully which aircraft would be best suited for scientific investigations, in terms of quality and duration of reduced gravity level but also ease of access and technical support from the integration team. In this respect, this section introduced the main aircraft used throughout the world to conduct research in reduced gravity. This survey is probably not exhaustive as no information could be obtained on similar parabolic flight program for reduced gravity research and practical training from emerging space-faring nations like China, India, Brazil, and possibly other countries.

References

[1] Clervoy, J.F., J.P. Haigneré, V. Pletser, and A. Gonfalone. "Microgravity with Caravelle." *Proceedings of the VIIth European Symposium Materials & Fluid Sciences in Microgravity*, ESA SP-295, Noordwijk, 685–691, 1990.

[2] Pletser, V., D. Thiérion, and B. Boissier, eds. "Experiment Results of ESA and CNES Campaigns, Tenth Anniversary of First ESA Parabolic Flight Campaign." *ESA WPP-90*, Noordwijk, 1995.

[3] Pletser, V. "Short Duration Microgravity Experiments in Physical and Life Sciences During Parabolic Flights: The First 30 ESA Campaigns". *Acta Astronautica* 55 (2004): 829–854.

[4] Pletser, V., A. Pacros, and O. Minster. "International Heat and Mass Transfer Experiments on the 48th ESA Parabolic Flight Campaign of March 2008." *Microgravity Science and Technology*, special Issue Two-Phase Systems 20 (2008): 177–182.

[5] Pletser, V., Pacros A., Minster O. "The ESA Parabolic Flight Programme for Physical Sciences." 3rd International Symposium Physical Sciences in Space (ISPS), Nara." *Journal Japan Society of Microgravity Application* 25-3 (2008): 635–640.

[6] Pletser, V., J. Winter, F. Duclos, U. Friedrich, J.F. Clervoy, T. Gharib, F. Gai, O. Minster, and P. Sundblad. "The First Joint European Partial-g Parabolic Flight Campaign at Moon and Mars Gravity Levels for Science and Exploration." *Microgravity Science and Technology*, special Issue ELGRA 2011, 24–6 (2012): 383–395.

[7] Gerathewohl, S.J. ed. "Zero-G Devices and Weightlessness Simulators." *Report for the Armed Forces—NAS-NRC Committee on Bioastronautics Panel on Acceleration*, National Academy of Sciences USA, National Research Council, Publication 781, 1961.

[8] "The History of KC-135A Reduced Gravity Research Program", NASA/JSC Aircraft Operations website, http://jsc-aircraft-ops.jsc.nasa.gov/Reduced_Gravity/KC_135_history.html, Last accessed 09/04/2013.

[9] Frimout D., and A. Gonfalone. "Parabolic Aircraft Flights, an Effective Tool in Preparing Microgravity Experiments." *ESA Bulletin* 42, Noordwijk (1985): 58–63.

[10] "NASA/JSC Aircraft Operations: The History of KC-135A Reduced Gravity Research Program." http://jsc-aircraft-ops.jsc.nasa.gov/Reduced_Gravity/KC_135_history.html. Last accessed 24/02/2013

[11] "About The C-9B Aircraft." NASA/JSC Aircraft Operations website, http://jsc-aircraft-ops.jsc.nasa.gov/Reduced_Gravity/about.html. Last accessed 24/02/2013

[12] "G-Force One, NASA/JSC Reduced Gravity Office with Zero Gravity Corporation." Nasa Flight Opportunities website, https://flightopportunities.nasa.gov/platforms/parabolic/gforce-one/, Last accessed 09/04/2013.

[13] "Zero-G Corporation: How it works." Zero-G Corporation website, http://www.gozerog.com/index.cfm?fuseaction=Experience.How_it_Works, http://www.gozerog.com/index.cfm?fuseaction=Research_Programs.welcome, Last accessed 24/02/2013

[14] "IL-76MD", Ilyushin website, http://www.ilyushin.org/eng/products/military/76md.html, last accessed 09/04/2013.

[15] Pletser, V., and C. Cornuejols. "Micro-accelerometric measurements on board the Russian Ilyushin IL-76 MDK during parabolic flights", *Proceedings of the Joint ESA-CNES Workshop Experiment Results of ESA and CNES Parabolic Flight Campaigns*, ESA WPP-090, 203–219, 1995.

[16] "Cessna Citation Laboratory Aircraft." TU Delft Aerospace Engineering website, http://www.lr.tudelft.nl/en/organisation/departments/control-and-operations/control-and-simulation/facilities-and-institutes/cessna-citation-laboratory-aircraft/, last accessed 09/04/2013.

[17] "Flight test and evaluation–Falcon 20." NRC website, http://www.nrc-cnrc.gc.ca/eng/solutions/advisory/flight_test.html, last accessed 09/04/2013.

[18] "Micro-Gravity Experiment." Diamond Air Service website, http://www.das.co.jp/new_html_e/service/01-1.html, last accessed 09/04/2013.

[19] Zeredo, J.L., K. Toda, M. Matsuura, Y. Kumei. "Behavioral Responses to Partial Gravity Conditions in Rats." *Neuroscience Letters* 529 (2012): 108–111.

[20] Zeredo, J.L., K. Toda, and Y. Kumei. "Neuronal Activity in the Subthalamic Cerebrovasodilator Area Under Partial-Gravity Conditions in Rats." *Life* 4 (2014): 107–116.

[21] Hasegawa, K., P.S. de Campos, J.L. Zeredo, Y. Kumei. "Cineradiographic Analysis of Mouse Postural Response to Alteration of Gravity and Jerk (gravity deceleration rate)." *Life* 4 (2014): 174–188.

[22] "EXA and FAE develops first Zero-G plane in Latin America." Agencia Espacial Civil Ecuatoriana website, http://exa.ec/bp16/index-en.html, last accessed 09/04/2013.

[23] "Astronautentraining/Schwerelosigkeit für Alle !!! ", Pauls Parabelflüge website, http://www.bierl.at/, last accessed 09/04/2013.

[24] Studer, M., G. Bradacs, A. Hilliger, E. Hurlimann, S. Engeli, C.S. Thiel, P. Zeitner, B. Denier, M. Binggeli, T. Syburra, M. Egli, F. Engelmann, and O. Ullrich. "Parabolic Maneuvers of the Swiss Air Force Fighter Jet F-5E as a Research Platform for Cell Culture Experiments in Microgravity." *Acta Astronautica* 68 (2011): 1729–1741.

[25] Pletser V., P. Queeckers, O. Dupont, J.C. Legros. "Parabolic Flights with the Fouga Magister aircraft." *Proceedings of the Joint ESA-CNES Workshop Experiment Results of ESA and CNES Parabolic Flight Campaigns*, ESA WPP-090, 221–231, 1995.

8

Magnetic Levitation

Clement Lorin[1], Richard J. A. Hill[2] and Alain Mailfert[3]

[1]Mechanical Engineering Department, University of Houston, Houston, TX, USA
[2]School of Physics and Astronomy, University of Nottingham, Nottingham, UK
[3]Laboratoire Géoressources, CNRS-Université de Lorraine, Nancy, France

8.1 Introduction

To compensate gravity magnetically, one generates at all points within an object—at the molecular scale—magnetic forces that counterbalance the force of gravity. This form of magnetic levitation, distinct from levitation of a ferromagnet or superconductor, or flotation, exploits the fact that a diamagnetic or paramagnetic medium is subjected to a magnetic body force in a non-uniform magnetic field. Suitable magnetic field sources for levitation include windings carrying continuous currents and in some cases permanent magnets. The first simulations of space conditions by levitation in fluids or biological systems were reported using water-cooled or superconducting magnets [1–4]. Magnetic compensation of gravity has many advantages compared to other ways to access weightlessness, such as unlimited experimentation time, ground-based experiments, and easily adjustable level of simulated gravity. It is the only way to directly compensate gravity at the molecular level, on the ground. Partial gravity compensation allows simulation of the gravity on Moon or Mars [5]. Here, we discuss the accuracy with which one can compensate gravity using magnetic levitation, for fluids and for biology.

8.2 Static Magnetic Forces in a Continuous Medium

8.2.1 Magnetic Forces and Gravity, Magneto-Gravitational Potential

Windings carrying electrical currents with a volume density $\boldsymbol{J}$ (A m^{-2}) or magnetized magnetic media [magnetization density $\boldsymbol{M}$ (A m^{-1})] generate a magnetic field of excitation $\boldsymbol{H}$ (A m^{-1}). Experimentally, one measures the magnetic flux density $\boldsymbol{B}$ (T or Wb m^{-2}). In free space, $\boldsymbol{B} = \mu_0 \boldsymbol{H}$ where $\mu_0 = 4\pi \times 10^{-7}$ H m^{-1} is the vacuum permeability. The relations between these quantities are, in stationary (time-independent) conditions, as follows:

$$\mathbf{curl}\,(\boldsymbol{H}) = \boldsymbol{J}; \quad \mathrm{div}\,(\boldsymbol{B}) = 0 \tag{8.1}$$

in a current-free domain:

$$\mathbf{curl}\,(\boldsymbol{H}) = \mathbf{0} \tag{8.2}$$

A magnetic field $\boldsymbol{H}$ applied to any homogeneous material (solid or fluid) produces a magnetization density:

$$\boldsymbol{M}\,(\boldsymbol{H}) = \chi \boldsymbol{H}. \tag{8.3}$$

The dimensionless magnetic susceptibility χ can be positive (paramagnetism) or negative (diamagnetism). We will take into account non-hysteretic media with χ considered as a constant when the magnetic field varies. That is neither ferromagnetic media nor colloidal suspensions of particles, namely ferrofluids.

In these media, the flux density is as follows: $\boldsymbol{B} = \mu_0\,(\boldsymbol{H} + \boldsymbol{M}) = \mu_0\,(1 + \chi)\boldsymbol{H}$, where $\mu_0(1 + \chi)$ is the medium magnetic permeability.

The above relations give unique solutions for $\boldsymbol{H}$ and $\boldsymbol{B}$, given distributions for $\boldsymbol{J}$ and $\boldsymbol{M}$. However, the "designers" of a levitation magnet have to solve the "inverse problem": Which configurations of $\boldsymbol{J}$ and $\boldsymbol{M}$ can create given distributions of $\boldsymbol{H}$ and $\boldsymbol{B}$ inside a finite 3D domain? This problem has numerous possible solutions.

A magnetic energy is associated with the magnetization that appears in materials. If the field $\boldsymbol{H}$ varies spatially, the magnetic energy varies spatially, too. A magnetic material is thus subjected to a volume force that can be written as follows:

$$f_{\mathbf{m}} = \frac{\mu_0}{2}\chi\,\mathbf{grad}(\boldsymbol{H}^2)$$

In materials that are considered for magnetic levitation, $|\chi| \ll 1$, that is $\mu_0\,(1 + \chi) \sim \mu_0$. Therefore, the magnetic volume force can likewise be expressed as follows:

$$\boldsymbol{f}_{\mathbf{m}} = \frac{1}{2\mu_0}\chi \boldsymbol{G} \tag{8.4}$$

Table 8.1 Order of magnitude of some fluid features for magnetic compensation of gravity

Fluid (P = 1 bar, g = 9.81 m s^{-2})	O_2 (90 K)	H_2 (20 K)	H_2O (293 K)	He (4.2 K)
Magnetic susceptibility (10^{-6})	+3,500	–1.8	–9.0	–0.74
Density (kg m^{-3})	1,140	71	998	125
G_1 (T^2 m^{-1})	+8	–990	–2,720	–4,140

where $\boldsymbol{G} = \mathbf{grad}\,(\boldsymbol{B}^2)$. The principle of magnetic gravity compensation is based on relation (8.4). The magnetic force must be opposite to the force applied by gravity $\boldsymbol{g}$ (m s^{-2}): $\boldsymbol{f}_{\mathbf{G}} = \rho \boldsymbol{g}$.

Therefore, the condition of magnetic compensation at any point within a medium of susceptibility χ and density ρ is

$$\boldsymbol{grad}\,(\boldsymbol{B}^2) = -2\mu_0 \frac{\rho}{\chi} \boldsymbol{g} = \boldsymbol{G}_1 \tag{8.5}$$

where (χ/ρ) (m^3 kg^{-1}) is the mass susceptibility. It is also useful to consider the total resulting potential energy Σ_L (J m^{-3}), named magneto-gravitational potential (MG potential), at a given height z (m) [3].

$$\Sigma_{\mathrm{L}} = \frac{\chi B^2}{2\mu_0} - \rho g z$$

From relation (8.5), Table 8.1 gives some algebraic values of $\boldsymbol{G}_1$ needed to compensate gravity for various fluids commonly used in aerospace engineering.

8.2.2 Magnetic Compensation Homogeneity

There is a fundamental constraint for magnetic compensation of gravity. A given $\boldsymbol{B}$ distribution leads to a unique force distribution, but an arbitrary force distribution cannot be obtained. Indeed, in a 3D domain, small with respect to the size of Earth, the terrestrial acceleration is independent of position, that is $\boldsymbol{G}$ should be uniform to satisfy (8.5). Since magnetic flux density $\boldsymbol{B}$ obeys (8.1) and (8.2), $\boldsymbol{G}$ cannot be simultaneously nonzero and uniform in the whole 3D domain [6]; it follows that it is impossible to obtain perfect compensation of gravity at every point in space! However, there exist magnetic field distributions where compensation is perfect in several points, or along a vertical segment line, or on a horizontal plane. It follows from the previous theorem that it is impossible to magnetically compensate any force field with a zero divergence. But it has been shown theoretically [7] that for diamagnetic materials, gravity can be compensated exactly, in the whole 3D domain with translational invariance, by combination of both magnetic and centrifugal forces $\boldsymbol{\Omega}$ (note than div ($\boldsymbol{\Omega}$) $\neq 0$). No experimental verification of this last property has been carried out yet. In the absence of $\boldsymbol{\Omega}$, we define a (dimensionless) inhomogeneity vector ε which represents the relative error between the $\boldsymbol{G}$ produced by the magnet and that required for perfect gravity compensation $\boldsymbol{G}_1$,

$$\varepsilon = \frac{\boldsymbol{G} - \boldsymbol{G}_1}{\overline{\boldsymbol{G}_1}}$$

Here, $\overline{\boldsymbol{G}_1}$ is the algebraic value of $\boldsymbol{G}_1$; that is, $\overline{\boldsymbol{G}_1}$ is positive for paramagnetic materials and negative for diamagnetic ones ($\overline{\boldsymbol{G}_1}$ is written G_1 up to the end of the section). The quality of the compensation can be directly visualized by mapping the vector field ε. Indeed, the latter represents the residual acceleration $\Gamma = g\varepsilon$, which can be interpreted as an effective gravitational field Γ. In order to use magnetic compensation method, one first has to define what approximation to the exact compensation is needed inside the working zone.

8.3 Axisymmetric Levitation Facilities

Levitation devices are usually made of axisymmetric windings. An analysis is first developed for a single vertical-axis solenoid and then extended to more complex systems.

8.3.1 Single Solenoids

Figure 8.1 shows the magnetic compensation obtained using a solenoid current I high enough to for the magnetic force to balance gravity at two points, one stable (+) and one unstable ($\times$), on the solenoid axis. In the right-hand panel, I is larger than in the left, showing that the axial location of the points (+, $\times$), the shape of the MG equipotentials (MGE), and the resulting acceleration ε depend on I.

Using Taylor polynomial approximations of the relevant quantities, one can establish relations between the components of the inhomogeneity vector (ε_r and ε_z in cylindrical coordinates) at a point a distance R from the perfect magnetic compensation point (+) and the norm B of the magnetic flux density at this point for a given fluid (G_1 constant). The first-order approximation in axisymmetric levitation facilities such as the above solenoid leads to

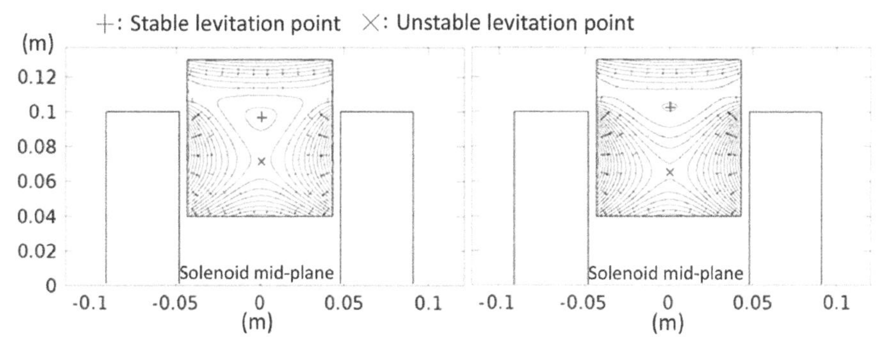

Figure 8.1 MGE (*blue curves*) and inhomogeneity ε (*black arrows*) in an arbitrary volume surrounding both points of perfect compensation along the solenoid axis, for two different currents. Dimensions are those of solenoid HyLDe used at the French Atomic Energy Commission (CEA Grenoble) for LH2. The stable point (*plus symbol*) is at the bottom of a local potential well, and the unstable point (*multipication symbol*) is a saddle point in the potential.

$$B = \frac{1}{2}\left[\frac{3\left|G_1\right| R}{2\varepsilon_{\mathrm{r}} + \varepsilon_{\mathrm{z}}}\right]^{\frac{1}{2}}$$

It is worth noticing that R/ε varies as the square of the magnetic flux density for a given fluid. Therefore, high magnetic field facilities are used to reach better magnetic compensation or larger levitated volumes. This expression shows that the levitation in a 1-mm-radius sphere, with a maximal isotropic homogeneity of 99 %, that is $\varepsilon_r = \varepsilon_z = 1$ %, requires a magnetic flux density of 5.0 T for hydrogen ($G_1 = -990$ T^2 m^{-1}) and 8.3 T for water ($G_1 = -2{,}740$ T^2 m^{-1}). But the shape of isohomogeneity zones near the compensation point is not necessarily spherical: They may be ellipsoidal, as it will be shown below. A spherical harmonic analysis of the magnetic field allows one to determine analytically the resulting acceleration around the perfect stable compensation point; the results are given in Figure 8.2, encapsulating the diamagnetic compensation performance of the superconducting single solenoid HyLDe. The C_n coefficient (C_1, C_2, or C_3 in Figure 8.2) is the nth spherical harmonic coefficient of the magnetic scalar potential W: $W_n = C_n r^n P_n(\cos\theta)$ where W_n is the nth harmonic of the magnetic scalar potential inside the magnetic field sources, r is the radius, θ is the polar angle of the spherical coordinate system, and P_n is the nth-degree Legendre polynomial. The magnetic field derives from this potential: $\boldsymbol{H} = \mathbf{grad}(-W)$. The C_n coefficients vary depending on the origin of the coordinate system along the axis as shown in Figure 8.2.

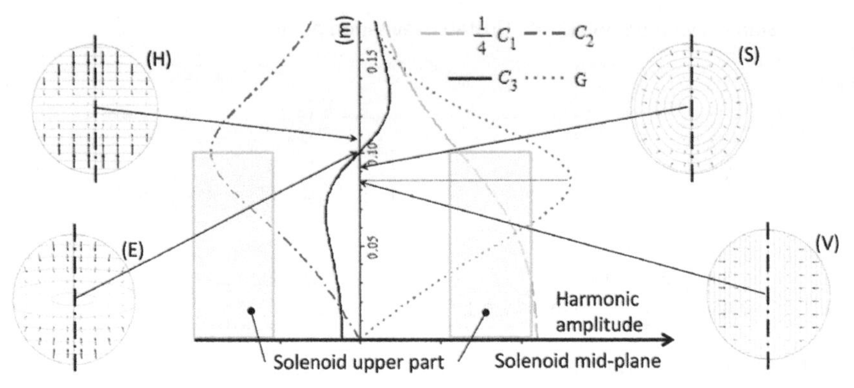

Figure 8.2 Variations of first three spherical harmonics (C_1, C_2, C_3) of the scalar potential W of the field, along the *upper part* of the axis of the solenoid (HyLDe) at a given current. The *red dotted line* is the amplitude of the vector $\boldsymbol{G}$ (proportional to C_1 times C_2). On the axis are located the first levitation point (V) and three other specific levitation points (S, E, H). The levitations occur, respectively, at z_V = 0.085 m, z_S = 0.092 m, z_E = 0.101 m, and z_H = 0.113 m at different current values I_V, I_S/I_V = 1.012, I_E/I_V = 1.060, and I_H/I_V = 1.111. The theoretical shapes of the MG potential wells surrounding the levitation points as well as the resulting acceleration (*black arrows*) are plotted.

The red dotted line gives, along the solenoid axis, the vertical component of the volume force $\boldsymbol{f}_m$, which is proportional to the square of the solenoid current, I. At a value $I = I_V$ of the current, the first levitation point is reached for liquid hydrogen (LH2), at a point V on the axis. For $I > I_V$, two levitation points exist along the axis; only the upper one is stable in the vertical direction. Increasing the current further, one successively goes through various MG potential well configurations around the successive stable levitation points. The well shapes are, respectively, prolate ellipsoids (between V and S), spheres (point S), oblate ellipsoids (between S and H), and horizontal planes (point H). The example here is for LH2, but the same is true of other liquids, such as water [8]. For a single solenoid, the residual forces around the stable compensation point can only be modified if the position of this point is changed and consequently the current too. Practically, there can be a technological limitation due to the maximal current value, resulting from the cooling of resistive magnets, or the critical surface of superconducting magnets. Since the spherical harmonic coefficients of the magnetic field and so the field $\boldsymbol{B}$ at any point in the working zone are uniquely related to the spatial derivatives of $\boldsymbol{B}$ along the axis, for a single solenoid as well as for any axisymmetric set of windings, the above analysis can be applied if $\boldsymbol{B}$ along the symmetry axis is known.

8.3.2 Improvement of Axisymmetric Device Performance

8.3.2.1 Ferromagnetic inserts

Magnetic force field distributions and thus MGE configurations can be changed by means of ferromagnetic inserts located close to the working zone. Figure 8.3 shows a ring-shaped insert that could be used to enlarge the 1 % inhomogeneity levitation zone (equivalent insert has been manufactured at CEA Grenoble, HyLDe facility). Elongation of the ellipsoidal MG well was obtained, using this insert. If the ferromagnetic insert becomes saturated, the norm of $\boldsymbol{G}$ (hence the magnetic force magnitude) is no longer proportional to the square of the solenoid current.

8.3.2.2 Multiple solenoid devices and special windings design

As of 1999, multiple solenoid devices were designed to increase field, magnetic forces, and improve magnetic compensation quality [9, 10], but without MGE tuning. A control of the different currents in multiple coil levitation devices should allow for continuous tuning of the MGE around perfect levitation points. As far as we know, the best device for MGE tuning should be a set of harmonic coils where each coil would be fed by independent current.

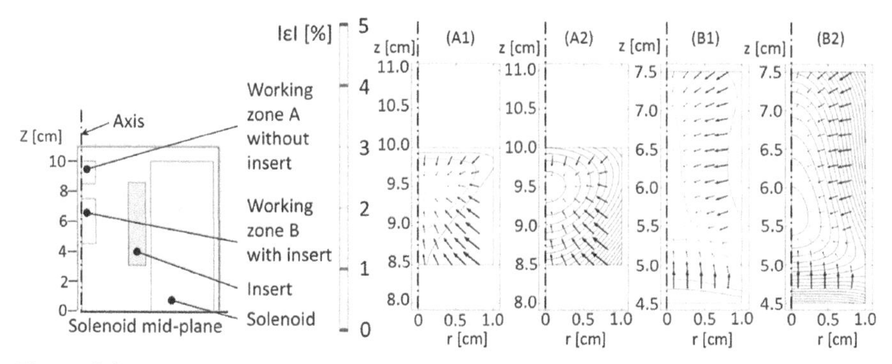

Figure 8.3 Comparison of magnetic compensation quality within the bore of a single solenoid with and without insert. On the *left* is an overview of the system. Adding an insert modifies the force configuration. The levitation points are changed as well as the current needed to reach the levitation. Thus, there are two different working zones: **A** (no insert, $J_A = 218.28$ A mm^{-2}) and **B** (insert, $J_B = 251.94$ A mm^{-2}). Working zone location and current are defined so as to get the largest levitated volume at given homogeneity. The resulting acceleration (*black arrows*) shows that levitation is stable inside both of the cells **A** and **B**. Isohomogeneity (*color curves*) is provided from 1 to 5 % by step of 1 % in figures **A1** and **B1**. MGE iso-Σ_L (*blue curves*) are elongated by the insert in the vertical direction as shown in figure **B2** w.r.t. figure **A2**.

This multicoil device allowing arbitrary control of the first three harmonics would enable quick variation of gravity by rapid current variation in one coil, as well as easy adjustment of residual forces [11]. Such a design has not been built yet. This approach leads to device design dedicated to the creation of useful magnetic force distributions for specific applications. For example, an axisymmetric distribution of $\boldsymbol{B}$ varying as $B_0\,(z - z_0)^{1/2}$ on the vertical axis of symmetry, generated by an appropriate current distribution in the windings, gives a constant norm of $\boldsymbol{G}$ along a vertical segment (but not in a whole 2D axisymmetric domain); here, B_0 and z_0 are constants. Besides axisymmetric coils, the development of saddle coils for particle accelerator magnets should allow for new types of levitators with a long working zone. The general theory for axisymmetric systems can be easily transposed to these magnets. It is worth noticing that the theoretical development presented here can be applied to permanent magnet systems. High-coercivity magnets are in any way equivalent to high value of surface current density (A m^{-1}), leading to high magnitude of $\boldsymbol{G}$ in narrow zones, near magnet edges. Such permanent magnet systems have been designed for free-contact handling of diamagnetic microdroplets (biomedical applications). Levitation is observed if the integrated magnetic volume force over the whole droplet volume counterbalances the droplet weight since the surface tension is very high at those dimensions (levitation similar to that of a rigid body); therefore, the quality of local compensation is not an issue.

8.4 Magnetic Gravity Compensation in Fluids

Many experiments in magnetic gravity compensation focus on fluid mechanics, especially on gas bubbles, liquid drops, and diphasic fluids with or without temperature gradient [3, 8, 12–18]. An important issue when levitating a drop or bubble is the stability of mechanical equilibrium. Drops and bubbles are stable when they are trapped within a well of the MG potential. Stable diamagnetic levitation of droplets inside a vertical-axis single solenoid is possible in the upper region of the solenoid (marked "+" on Figure 8.1). Conversely, paramagnetic levitation of bubbles occurs in the lower part of the solenoid, and only the lowest levitation point is stable in the vertical direction. In stable mechanical equilibrium, the center of a vanishingly small and spherical bubble, or droplet, is located at the local minimum of the MG potential Σ_L. For finite-sized droplets or bubbles, both its shape and its equilibrium position depend on various parameters: surface tension of the liquid/gas interface, the MG potential well shape, and demagnetization energy. The latter is due to the magnetic field modification because the gas/liquid interface shape always tends to elongate paramagnetic bubbles or diamagnetic droplets in the magnetic field direction. This effect has been revealed [13] and then analyzed [14] in paramagnetic liquid oxygen. For a diamagnetic gas/liquid interface, this effect can be neglected due to the weakness of the magnetic susceptibility. Interface shape is then given by equilibrium of surface tension and MG potential [8, 15]. When diamagnetic liquids are close enough to their critical point, the interface shape tends to that of the MGE [18]. In paramagnetic fluids, patterning effects of the liquid/gas interface may be observed when the magnetic field is perpendicular to the interface [19]. This effect, well known for ferrofluids, can make levitation useless in paramagnetic diphasic systems such as liquid/gas oxygen. Thermal exchanges have been studied in various boiling regimes [13]. Observed phenomena, bubble size and growth rate, are far different from those under normal gravity but rather similar to what happens in real weightlessness. Magneto-convection can appear in paramagnetic monophasic fluid due to the susceptibility dependency on the temperature. This effect has been observed and investigated by means of a magnetic Rayleigh number [20]. All of these examples of bubble shape stability, gas/liquid interface modifications, and thermal convection demonstrate that the microgravity environment generated by magnetic field can, in many ways, closely mimic true weightlessness. However, the strong magnetic field can produce other effects not observed in weightlessness. Some of these effects can be mitigated by magnet design, through the harmonic contents of the field.

8.5 Magnetic Gravity Compensation in Biology

Since Geim and coworkers [2] demonstrated levitation of a live frog and Valles et al. [4] studied a levitating frog's egg in 1997, diamagnetic levitation has been used in experiments on a variety of biological organisms, including, for example, *Paramecia* [21], yeast [22, 23], *Arabidopsis* plants [23–25] and cell cultures [26–28], bacteria [29–31], bone cells [32–36], a live mouse [37], and fruit flies [38, 39].

Since the diamagnetic force is a body force, like the gravitational force, we can write the net force acting on the object in the magnetic field in terms of an *effective* gravitational field, $\Gamma = \boldsymbol{g}\varepsilon = \boldsymbol{g} + (\chi/\rho)\boldsymbol{G}/(2\mu_0)$; ε, $\boldsymbol{g}$, and $\boldsymbol{G}$ are as defined above. The ratio χ/ρ is constant for a homogeneous substance such as water or a well-mixed solution. Spatial variations in Γ owing to spatial variation of the magnetic field exert weak tidal forces on an object in the magnetic field: variation in Γ is typically $\sim$0.5 m s^{-2} per cm in current experiments [8], but can be made as small as one like along a line segment using a suitable solenoid design [40].

For biological material, χ/ρ cannot be taken as a constant: It varies depending on the tissue type. For most soft biological tissues, $-11.0 \times 10^{-6} < \chi < -7.0 \times 10^{-6}$ [41], close to the value for water, -9×10^{-6}. Iron-rich tissues have slightly more positive susceptibility owing to the paramagnetism of the Fe ion. However, even an iron-rich organ such as liver has susceptibility only slightly more positive than water. Table 8.2 gives the susceptibilities of a few biological materials, along with approximate densities ρ. Variations in χ/ρ between different material types give rise to variations in the effective gravity Γ throughout the organism, generating physical stresses akin to the above-mentioned tidal forces. Values of $\varepsilon = \Gamma/g$ at the stable levitation point of water are given in the table, to enable comparison. At the levitation point of the *organism*, determined by its mean susceptibility (usually close to that of water) and density, the magnitude of Γ may be slightly smaller: Experiments on freely levitating frogs' eggs determined the magnitude of Γ acting on three main constituents of the egg fractioned by centrifugation—the cytosol, a protein-rich pellet, and lipids—to be 0.02 g, 0.075 g, and 0.06 g, respectively [4].

Table 8.2 suggests that relatively dense materials, such as bone and starch, will experience a significant fraction of g even in a freely levitating organism. Substantiating this, experiments show that a growing plant root can readily establish the direction of real gravity when levitating [23, 25], consistent with the theory that plants use the position of starch-rich *statoliths* within the cells at the root tip to sense which way is down. The buoyancy of a

Table 8.2 Volume susceptibility χ and approximate density ρ of some biological materials and tissues, including the magnitude of the effective gravity $\varepsilon = |\Gamma|/g$ and its direction (up or down with respect to gravity, indicated by arrows), calculated at the levitation point of water

	χ (10^{-6})	ρ (kg m^{-3})	ε (at lev. pt. water)
Water (37 °C)	−9.05 [41]	993 [41]	0
Cytosol of a frog's egg	−9.09 [4]	1,030 [4]	0.03↓
Stearic acid (20 °C)	−10.0 [41]	940 [42]	0.18↑
Starch	−10.1 [43]	1,530 [42]	0.27↓
Whole blood, human (deoxyg.)	−7.90 [41]	1,040 [44]	0.16↓
Liver, human (healthy)	−8.8 [41]	1,050 [41]	0.08↓
Cortical bone, human	−8.9 [41]	1,900 [44]	0.49↓
Cholesterol	−7.61 [45]	1,020 [45]	0.18↓

statolith within a root cell is altered by the field gradient $\boldsymbol{G}$, through the magneto-Archimedes effect [19, 46]. The magnitude of $\boldsymbol{G}$ required to keep a statolith *neutrally buoyant* within the cell is 3–4 times as large as that required to levitate water [23]. Such differences in $\boldsymbol{G}$ required for flotation can also be exploited for separation of biological [45] and non-biological materials [19]. Magneto-Archimedes buoyancy can also be responsible for magneto-convection in liquid microbiological cultures [30].

In order to differentiate between effects of levitation and any other effects of the strong magnetic field, at least two chambers containing the biological samples are usually placed in the magnetic field, one enclosing the point where the sample levitates and another enclosing the geometric center of the solenoid, where $\boldsymbol{G} = \boldsymbol{0}$. The central chamber is used to control for other effects of magnetic field, besides levitation. Chambers may be placed at other points in the field to simulate Martian or lunar gravity for example [5], or to simulate hypergravity. The confinement of the chamber may be used to restrict the range of effective gravities to which the organism is exposed and is made as small as required for this purpose. Comparing levitating samples with those in 2 g hypergravity can reveal the relative influence of stresses induced by differences between Γ acting on different tissue types; if such stresses dominate, results from levitation and 2 g would be expected to be similar [39]. Using this technique, the movements of levitating fruit flies were found to be consistent with those observed in true weightlessness, with no other effects of the strong magnetic field observed [39]. In other experiments, effects of the strong field ($\sim$10 T) are observed, besides that of levitation; see, for example, Refs. [31, 38, 47, 48].

The evidence emerging from experiments on a variety of different organisms suggests that levitation can compensate gravity quite effectively. In studies on frogs' eggs, for example, the authors find [4], "the reduction in body forces and gravitational stresses achieved with magnetic field gradient levitation... has not been matched by any other ground-based, low-gravity simulation technique." One should be aware, however, of the variation in effective gravity between different tissue types, particularly for higher-density materials such as bone, and that the strong magnetic field may influence the organism through other mechanisms, which may mask the effect of altered gravity.

Acknowledgments

We would like to thank the SBT-CEA-Grenoble for data about HyLDe facility.

References

[1] Beaugnon, E. and R. Tournier. "Levitation of Organic Materials." *Nature* 349(1991): 470.

[2] Berry, M.V. and A.K. Geim. "Of Flying Frogs and Levitrons." *European Journal of Physics*. 18 (1997): 307–313.

[3] Weilert, M.A., D.L. Whitaker, H.J. Maris and G.M. Seidel. "Magnetic Levitation and Noncoalescence of Liquid Helium." *Physical Review Letters*. 77 (1996): 4840 and "Magnetic Levitation of Liquid Helium." *Journal of Low Temperature Physics*, 106 (1997): 101–131.

[4] Valles, J.M., Jr., J.M. Denegre and K.L. Mowry. "Stable Magnetic Field Gradient Levitation of *Xenopus laevis*: Toward Low-Gravity Simulation." *Biophysical Journal* 731130-1133 (1997): 1130–1133.

[5] Valles, J.M., Jr., H.J. Maris, G.M. Seidel, J. Tang and W. Yao. "Magnetic Levitation-Based Martian and Lunar Gravity Simulator". *Advances in Space Research* 36 (2005): 114–118.

[6] Quettier, L., et al. "Magnetic Compensation of Gravity in Liquid/Gas Mixtures: Surpassing Intrinsic Limitations of a Superconducting Magnet by Using Ferromagnetic Inserts." *European Physical Journal Applied Physics* 32, no. 3 (2005): 167–175.

[7] Lorin, C. and A. Mailfert. "Magnetic Compensation of Gravity and Centrifugal Forces." *Microgravity Science and Technology* 21 (2009): 123–127.

[8] Hill, R.J.A. and L. Eaves. "Vibrations of a Diamagnetically Levitated Water Droplet." *Physical Review E* 81(2010): 056312; ibid. 85 (2012): 017301.

[9] Bird, M.D. and Y.M. Eyssa. "Special Purpose High Field Resistive Magnets." *IEEE Transactions on Applied Superconductivity* 10 (2000): 451–454.

[10] Ozaki, O., et al.. "Design Study of Superconducting Magnets for Uniform and High Magnetic Force Field Generation." *IEEE Transactions on Applied Superconductivity* 11 (2001): 2252–2255.

[11] Lorin, C. and A. Mailfert. "Design of a Large Oxygen Magnetic Levitation Facility." *Microgravity Science and Technology* 22 (2010): 71–77.

[12] Nikolayev, V., D. Chatain, D. Beysens and G. Pichavant. "Magnetic Gravity Compensation." *Microgravity Science and Technology* 23(2011): 113–122.

[13] Pichavant, G., B. Cariteau, D. Chatain, V. Nikolayev and D. Beysens. "Magnetic Compensation of Gravity: Experiments with Oxygen." *Microgravity Science and Technology* 21 (2009): 129–133.
[14] Duplat, J. and A. Mailfert. "On the Bubble Shape in a Magnetically Compensated Gravity Environment." *Journal of Fluid Mechanics* 716 (2013): R11.
[15] Chatain, D. and V. Nikolayev. "Using Magnetic Levitation to Produce Cryogenic Targets for Inertial Fusion Energy: Experiment and Theory." *Cryogenics* 42 (2002): 253–261.
[16] Beaugnon, E., D. Fabregue, D. Billy, J. Nappa and R. Tournier. "Dynamics of Magnetically Levitated Droplets." *Physica B* 294–295 (2001): 715–720.
[17] Hill, R.J.A. and L. Eaves. "Nonaxisymmetric Shapes of a Magnetically Levitated and Spinning Water Droplet." *Physical Review Letters* 101 (2008): 234501.
[18] Lorin, C., et al. "Magnetogravitational Potential Revealed Near a Liquid–Vapor Critical Point." *Journal of Applied Physics* 106 (2009): 033905.
[19] Catherall, A.T., L. Eaves, P.J. King and S.R. Booth. "Floating Gold in Cryogenic Oxygen." *Nature* 422 (2003): 579.
[20] Braithwaite, D., E. Beaugnon and R. Tournier. "Magnetically Controlled Convection in a Paramagnetic Fluid." *Nature* 354 (1991): 134–136.
[21] Guevorkian, K. and J.M. Valles, Jr. "Swimming *Paramecium* in Magnetically Simulated Enhanced, Reduced, and Inverted Gravity Environments."" *Proceedings of the National Academy of Sciences of the United States of America* 35 (2006): 13051–13056.
[22] Coleman, C.B., et al. "Diamagnetic Levitation Changes Growth, Cell Cycle, and Gene Expression of *Saccharomyces cerevisiae*." *Biotechnology and Bioengineering* 98 (2007): 854–863.
[23] Larkin, O.J. "Diamagnetic Levitation: Exploring the Effects of Weightlessness on Living Organisms." Ph.D. Thesis, University of Nottingham (2010).
[24] Brooks J.S., et al. "New Opportunities in Science, Materials, and Biological Systems in the Low-Gravity (magnetic levitation) Environment (invited)." *Journal of Applied Physics* 87 (2000): 6194–6199.
[25] Herranz, R., et al. "Ground-Based Facilities for Simulation of Microgravity: Organism-Specific Recommendations for their Use, and Recommended Terminology." *Astrobiology* 13 (2012): 1–17.
[26] Babbick, M., et al. "Expression of Transcription Factors After Short-Term Exposure of *Arabidopsis thaliana* Cell Cultures to Hypergravity

and Simulated Microgravity (2-D/3-D Clinorotation, Magnetic Levitation)." *Advances in Space Research* 39 (2007): 1182–1189.

[27] Manzano, A.I., et al. "Gravitational and Magnetic Field Variations Synergize to Cause Subtle Variations in the Global Transcriptional State of *Arabidopsis* In Vitro Callus Cultures." *BMC Genomics* 13 (2012): 105.

[28] Herranz, R., A.I. Manzano, J.J.W.A van Loon, P.C.M. Christianen and F.J. Medina. "Proteomic Signature of *Arabidopsis* Cell Cultures Exposed to Magnetically Induced Hyper- and Microgravity Environments." *Astrobiology* (2013). doi:10.1089/ast.2012.0883 (online before print).

[29] Beuls, E., R. Van Houdt, N. Leys, C.E. Dijkstra, O.J. Larkin and J. Mahillon. "*Bacillus thuringiensis* Conjugation in Simulated Microgravity." *Astrobiology* 9 (2009): 797–805.

[30] Dijkstra, C.E., et al. "Diamagnetic Levitation Enhances Growth of Liquid Bacterial Cultures by Increasing Oxygen Availability." *Journal of the Royal Society Interface* 6 (2011): 334–344.

[31] Liu, M., et al. "Magnetic Field is the Dominant Factor to Induce the Response of *Streptomyces avermitilis* in Altered Gravity Simulated by Diamagnetic Levitation." *PLoS One* 6 (2011): e24697.

[32] Hammer, B.E., L.S. Kidder, P.C. Williams and W.W. Xu. "Magnetic Levitation of MC3T3 Osteoblast Cells as a Ground-Based Simulation of Microgravity". *Microgravity Science and Technology* 21(2009): 311–318.

[33] Qian, A., et al. "cDNA Microarray Reveals the Alterations of Cytoskeleton-Related Genes in Osteoblast Under High Magneto-Gravitational Environment." *Acta Biochimica et Biophysica Sinica (Shanghai)* 41 (2009): 561–577.

[34] Qian, A.R., et al. "High Magnetic Gradient Environment Causes Alterations of Cytoskeleton and Cytoskeleton-Associated Genes in Human Osteoblasts Cultured In Vitro." *Advances in Space Research* 46 (2010): 687–700.

[35] Wang, L., et al. "Diamagnetic Levitation Cases Changes in the Morphology, Cytoskeleton, and Focal Adhesion Proteins Expression in Osteocytes."' *IEEE Transactions on Biomedical Engineering* 59 (2012): 68–77.

[36] Qian, A.-R., et al. "Large Gradient High Magnetic Fields Affect Osteoblast Ultrastructure and Function by Disrupting Collagen I or Fibronectin/$\alpha\beta$1 Integrin." *PLoS One* 8e51036 (2013): e51036.

[37] Liu, Y., D.-M. Zhu, D. M. Strayer and U.E. Israelsson. "Magnetic Levitation of Large Water Droplets and Mice." *Advances in Space Research* 45 (2010): 208–213.

[38] Herranz, R., et al. "Microgravity Simulation by Diamagnetic Levitation: Effects of a Strong Gradient Magnetic Field on the Transcriptional Profile of *Drosophila melanogaster*." *BMC Genomics* 13 (2012): 52.

[39] Hill, R.J.A., et al. "Effect of Magnetically Simulated Zero-Gravity and Enhanced Gravity on the Walk of the Common Fruit Fly." *Journal of the Royal Society Interface* 9 (2012): 1438–1449.

[40] Lorin, C., A. Mailfert, C. Jeandey and P.J. Masson. "Perfect Magnetic Compensation of Gravity Along a Vertical Axis." *Journal of Applied Physics* 113 (2013).

[41] Schenck, J.F. "The Role of Magnetic Susceptibility in Magnetic Resonance Imaging: MRI Magnetic Compatibility of the First and Second Kinds." *Medical Physics* 23 (1996): 815–850 (and references therein).

[42] Haynes, W.M. (ed.), *Handbook of Chemistry and Physics. 93rd edition.* Boca Raton, FL, CRC Press, 2012.

[43] Kuznetsov, O.A. and K.H. Hasenstein. "Intracellular Magnetophoresis of Amyloplasts and Induction of Root Curvature." *Planta* 198 (1996): 87–94.

[44] Cameron, J.R., J.G. Skofronick and R.M. Grant. "Physics of the Body". Madison WI, Medical Physics Publishing, 1992.

[45] Hirota, N., et al. "Magneto-Archimedes Separation and its Application to the Separation of Biological Materials." *Physica B* 346-347 (2004): 267–271.

[46] Ikezoe, Y., et al. "Making Water Levitate,." *Nature* 393 (1998): 749.

[47] Denegre, J.M., J.M. Valles Jr., K. Lin, W.B. Jordan and K.L. Mowry. "Cleavage Planes in Frog Eggs are Altered by Strong Magnetic Fields." *Proceedings of the National Academy of Sciences of the United States of America* 95 (1998): 14729–14732.

[48] Iwasaka, M., J. Miyakoshi and S. Ueno. "Magnetic Field Effects on Assembly Pattern of Smooth Muscle Cells." *In Vitro Cellular & Developmental Biology—Animal* 39 (2003): 120–123.

9

Electric Fields

Birgit Futterer[1,2], Harunori Yoshikawa[3], Innocent Mutabazi[3] and Christoph Egbers[1]

[1]Brandenburg University of Technology, Cottbus, Germany
[2]Otto von Guericke Universität, Magdeburg, Germany
[3]LOMC, UMR 6294, CNRS-Université du Havre, Le Havre, France

9.1 Convection Analog in Microgravity

Thermal convection within fluids is ubiquitous in nature and engineering. It plays a major role in heat transfer and is a main driver for geophysical and atmospheric structures. This thermally driven convection is conjoined with Archimedean buoyancy force due to the variation of the density with the temperature T in the gravitational field $\vec{g}$. In most of the fluids, the density decreases with the temperature and its behavior can be modeled by a linear relation for a small temperature variation: $\rho(T) = \rho_0[1 - \alpha(T - T_{ref})]$, where $\rho_0 = \rho\ (T_{ref})$, T_{ref} is the reference temperature, and a is the volume thermal expansion coefficient. The Archimedean buoyancy force reads

$$\vec{F} = -\rho_0\, \alpha\ (T - T_{ref})\ \vec{g} \tag{9.1}$$

Thermal convection induced by Archimedean buoyancy in the fluid layer confined between two parallel horizontal plates has been widely investigated since long time and is known as Rayleigh-Bénard convection [1]. It develops with a critical wave number $q_c = 3.117$, when the Rayleigh number $Ra = \alpha\Delta T g d^3/\nu\kappa$ exceeds its critical value Ra = 1707.8 (ΔT: temperature difference between the plates, ν and κ: kinematic viscosity and thermal diffusivity of a fluid, and d: the gap between the plates).

When gravity is absent, that is in microgravity conditions, may occur no phenomena related to the Archimedean buoyancy. However, it is possible to provoke thermal convection by using an electric field coupled with a temperature gradient applied to a fluid. This convection is often referred to as thermo-electrohydrodynamic (TEHD) convection. In fact, a dielectric fluid in the electric field $\vec{E}$ pertains to a ponderomotive force, the density of which is given by [2]

$$\vec{F} = \rho_f \vec{E} - \frac{1}{2}\vec{E}^2\vec{\nabla}\varepsilon + \vec{\nabla}\left[\rho\left(\frac{\partial\varepsilon}{\partial\rho}\right)_T \frac{\vec{E}^2}{2}\right] \tag{9.2}$$

where ρ_f is the free charge density. The first term is the Coulomb force density, the second term is called dielectrophoretic (DEP) force density, and the last one is the electrostriction force density. In case of incompressible fluid motion without interface, the last term can be lumped into the pressure term of the momentum equation. Thermo-electrohydrodynamics has been used as an active method for heat transfer enhancement [3, 4].

9.1.1 Conditions of DEP Force Domination

The spatial distribution of free charges varies under an electric field. This variation process occurs with a timescale $\tau_e = \varepsilon/\sigma$ called the charge relaxation time, where σ is the electric conductivity of the fluid. In dc electric field or ac electric field with a frequency $f < {\tau_e}^{-1}$, free charges accumulate at locations where σ varies in space, for example, at the surface of the fluid, and the Coulomb force density is often dominant component in (9.2). When the electric field is alternating at frequency $f >> {\tau_e}^{-1}$, then no free charge accumulation occurs. If the frequency is also higher than the inverse of the viscous relaxation timescale $\tau_\nu = d^2/\nu$, only the time-averaged components of (9.2) are concerned with the electrohydrodynamics so that the Coulomb force has no influence on it. Then, the DEP force, which always contains a static component, drives the electrohydrodynamics. For electric field frequency $f = 50\ Hz$, the relaxation times τ_e and τ_ν should be larger than 0.02 s.

Assuming the linear variation of the dielectric permittivity with temperature, that is $\varepsilon(T) = \varepsilon_{ref}[1 - \alpha_e(T - T_{ref})]$, the dielectrophoretic force can be reduced, after removing a gradient force component, to

$$\vec{F} = -\rho_0\,\alpha_e\,(T - T_{ref})\,\vec{g}_e \tag{9.3}$$

where we have introduced the electric gravity given by [5, 6]

$$\vec{g}_e = \nabla\left[\frac{\varepsilon_{ref}\alpha_e\vec{E}^2}{2\rho_0\alpha}\right]. \tag{9.4}$$

The electric gravity represents the gradient of the electrostatic energy stored in the dielectric fluid.

This TEHD convection driven by the DEP force in microgravity is the subject of this chapter. The chapter is organized as follows: After introducing the physical basis of the TEHD convection, we will discuss the electric gravity in three classic shapes of capacitors and then equations governing the convection development from the quiescent conductive state of the fluid. The chapter will end on some results from stability analysis and open questions and, additionally, will give a short summary on application in extended microgravity experiments.

9.1.2 Equations Governing DEP-Driven TEHD Convection

We consider a dielectric fluid confined inside a capacitor with applied alternating voltage $V(t) = \sqrt{2}\, V_0 \cdot \sin(2\pi ft)$. The TEHD convection in microgravity conditions may be described by mass and momentum equations coupled to energy and electric field equations. The assumption $f >> {\tau_e}^{-1}, {\tau_\nu}^{-1}$ allows for use of the time-averaged description; that is, the electric field and, hence, the electric gravity can be averaged over a period in the governing equations. In the Boussinesq approximation, the equations for TEHD convection read

$$\vec{\nabla}.\vec{u} = 0 \tag{9.5}$$

$$\frac{\partial \vec{u}}{\partial t} + \left(\vec{u}.\vec{\nabla}\right)\vec{u} = -\vec{\nabla}\Pi + \nu\vec{\nabla}^2\vec{u} - \alpha\left(T - T_{ref}\right)\vec{g}_e \tag{9.6}$$

$$\frac{\partial T}{\partial t} + \left(\vec{u}.\vec{\nabla}\right)T = \kappa\vec{\nabla}^2 T \tag{9.7}$$

$$\vec{\nabla}\cdot\left[\varepsilon(T)\vec{\nabla}\varphi\right] = 0 \text{ with } \vec{E} = -\vec{\nabla}\varphi \tag{9.8}$$

where φ is the electrostatic potential.

In the energy Equation (9.7), the viscous dissipation and Joule heating have been neglected, following the arguments developed by [7]. The reduced pressure Π is given by

$$\Pi = \frac{p}{\rho_0} - \frac{\alpha_e \epsilon_{ref}\left(T - T_{ref}\right)\vec{E}^2}{2\rho_0} - \left(\frac{\partial\varepsilon}{\partial\rho}\right)_T \frac{\vec{E}^2}{2}. \tag{9.9}$$

These equations must be solved together with appropriate boundary conditions at surfaces S_i of electrodes ($i = 1, 2$):

$$\vec{u} = 0 \quad T = T_1; \ \phi = V_0 \quad at \ S_1, \tag{9.10}$$

$$\vec{u} = 0 \quad T = T_2; \ \phi = 0 \quad at \ S_2. \tag{9.11}$$

From now on, we will consider T_2 as the reference temperature, that is $T_{ref} = T_2$, and correspondingly, ε_{ref} will be referred to as $\varepsilon_2 = \varepsilon\ (T = T_2)$.

9.2 Electric Gravity in the Conductive State for Simple Capacitors

Consider a dielectric fluid at rest between electrodes in simple geometrical configurations, that is plane, cylindrical, or spherical, with a temperature difference $\Delta T = T_1 - T_2$ and an alternative tension V_0 between the electrodes S_1 and S_2 (Figure 9.1). The temperature and electric fields can be computed analytically from the Equations (9.5–9.8), whereby the electric gravity can be derived by (9.4). Table 9.1 gives the expressions for these three configurations.

In plane capacitor, the gravity is due to the thermoelectric coupling through the thermoelectric parameter $B = \alpha_e \Delta T$; it is always oriented along the temperature gradient, that is $\vec{g}_e \uparrow\uparrow \vec{\nabla} T$. In cylindrical and spherical capacitors, the gravity is a product of two contributing factors: The first g_0 ($\sim r^{-n}$, $n = 3$ for cylindrical annulus and $n = 5$ for spherical shell) comes from the inhomogeneity of the electric field due to the curvature, and the second $F(B, \eta, r)$ is the thermoelectric coupling. Moreover, the electric gravity can be either centripetal, that is $\vec{g}_e \uparrow\downarrow \vec{e}_r$, or centrifugal

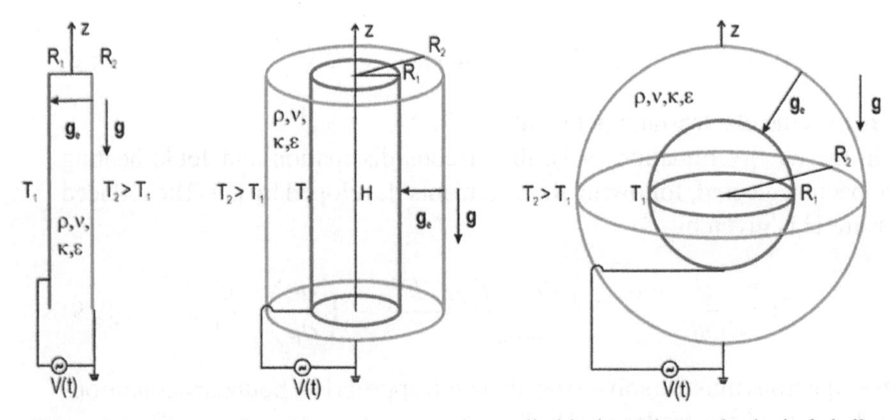

Figure 9.1 Flow configurations: plane capacitor, cylindrical annulus, and spherical shell.

Table 9.1 Basic conductive states in different electrode configurations. Parameter $B = \alpha_E \Delta T$ has been introduced

Capacitor Shape	Temperature Field	Electric Field	Electric Gravity
Plane capacitor	$T(x) = T_2 + \left(1 - \frac{x}{d}\right)\Delta T$	$\vec{E}(x) = E(x)\,\vec{e}_x;$ $E(x) = -E_2\left[1 - B\left(1 - \frac{x}{d}\right)\right]^{-1}$ $E_2 = \frac{B}{\ln(1-B)}\frac{V_0}{d}$	$\vec{g}(x) = -g_0\,F(B,x)\,\vec{e}_x;\ g_0 = \frac{\varepsilon_2 \alpha_E B}{\rho_0 \alpha d}\left(\frac{V_0}{d}\right)^2$ $F = \left[\frac{B}{\ln(1-B)}\right]^2 \frac{1}{[1-B(1-x/d)]^3}$
Cylindrical annulus	$T(r) = T_2 + \frac{\ln(r/R_2)}{\ln \eta}\Delta T$	$\vec{E}(r) = E(r)\,\vec{e}_r;$ $E(r) = -E_2\left[1 - B\frac{\ln(r/R_2)}{\ln \eta}\right]^{-1}\frac{R_2}{r}$ $E_2 = -\frac{B}{\ln(1-B)}\frac{V_0}{R_2 \ln \eta}$	$\vec{g}(r) = -g_0\,F(B,\eta,r)\,\vec{e}_r;\ g_0 = \frac{\varepsilon_2 \alpha_E}{\rho_0 \alpha (\ln \eta)^2}\frac{V_0^2}{r^3}$ $F = \left[\frac{B}{\ln(1+B)}\right]^2 \frac{1-(B/\ln \eta)[1+\ln(r/R_2)]}{[1-(B/\ln \eta)\ln(r/R_2)]^3}$
Spherical shell	$T(r) = T_2 + \frac{\eta}{1-\eta}\left(\frac{R_2}{r} - 1\right)\Delta T$	$\vec{E}(r) = E(r)\,\vec{e}_r;$ $E(r) = E_2\left(\frac{R_2}{r}\right)^2\left[1 - \frac{B\eta}{1-\eta}\left(\frac{R_2}{r} - 1\right)\right]^{-1}$ $E_2 = -\frac{B}{\ln(1-B)}\frac{\eta}{1-\eta}\frac{V_0}{R_2}$	$\vec{g}(r) = -g_0\,F(B,\eta,r)\,\vec{e}_r;$ $g_0 = \frac{2\varepsilon_2 \alpha_E}{\rho_0 \alpha}\left(\frac{\eta}{1-\eta}\right)^2 \frac{V_0^2 R_2^2}{r^5}$ $F = \left[\frac{B}{\ln(1-B)}\right]^2$ $\left[1 - \left(\frac{R_2}{2r} - 1\right)\frac{B\eta}{1-\eta}\right] \Big/ \left[1 - \left(\frac{R_2}{r} - 1\right)\frac{B\eta}{1-\eta}\right]^3$

$\vec{g}_e \uparrow\uparrow \vec{e}_r$, depending upon the sign of the function $F(B, \eta, r)$. The orientation of the basic electric gravity in the spherical shell is summarized in Figure 9.2. A detailed discussion of the electric gravity in cylindrical annulus can be found in [6].

9.2.1 Linear Stability Equations and Kinetic Energy Equation

The characteristic scales can be used to introduce non-dimensional control parameters. For timescale, we chose the viscous relaxation time τ_ν, the gap d between the electrodes is chosen as length scale, and ΔT is the temperature scale. The resulting control parameters are the Prandtl number $Pr = \nu/\kappa$ and the electric Rayleigh number $L = \alpha_e \Delta T g_m d^3/\nu k$, where g_m is the electric gravity at the mid-gap and the thermoelectric parameter B.

The linearized equations near the quiescent conducting state ($\vec{u} = 0$) are as follows:

$$\vec{\nabla}.\vec{u} = 0 \tag{9.12}$$

$$\frac{\partial \vec{u}}{\partial t} + \left(\vec{u}.\vec{\nabla}\right)\vec{u} = -\vec{\nabla}\Pi + \vec{\nabla}^2\vec{u} - \frac{L}{\mathrm{Pr}}\left[(T - T_2)\,\vec{g}_e' + \theta\vec{g}_e\right] \tag{9.13}$$

$$\mathrm{Pr}\left[\frac{\partial \theta}{\partial t} + \left(\vec{u}.\vec{\nabla}\right)T\right] = \vec{\nabla}^2\theta \tag{9.14}$$

$$\vec{\nabla}\left\{[1 - B(T - T_2)]\,\vec{\nabla}\phi - B\,\theta\vec{\nabla}\varphi\right\} = 0 \tag{9.15}$$

where ϑ and ϕ denote the perturbation temperature and electric potential, respectively. Two components have been distinguished in the electric

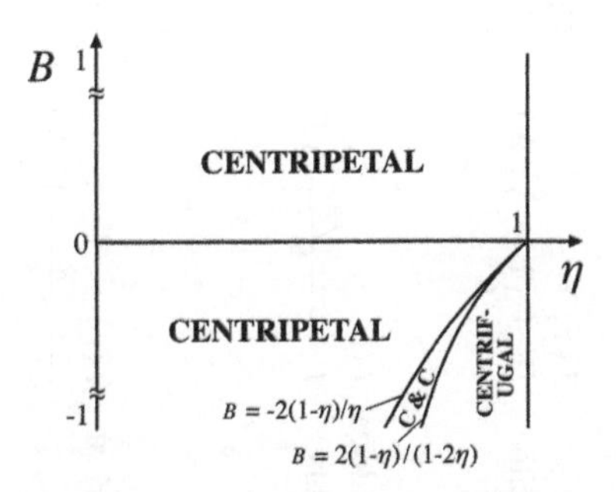

Figure 9.2 Diagram of basic gravity orientation in the spherical shell. C & C means that the gravity is centripetal and centrifugal in the inner and the outer layers, respectively.

gravity: $\vec{g}_e$ in the basic conductive state and $\vec{g}'_e$ related to the perturbation electric field. The latter arises through the thermoelectric coupling (9.15).

The equation of the perturbation kinetic energy is obtained straightforward from the previous equations and reads [6, 8]:

$$\frac{dK}{dt} = W_{BC} + W_{PC} - D_v \tag{9.16}$$

where

$$K = \int_V \frac{\vec{u}^2}{2} dV; \ \Pr W_{BG} = -L \int_V \theta \vec{u}.\vec{g}_e dV;$$

$$\Pr W_{PG} = -L \int_V \left[(T - T_2)\, \vec{u}.\vec{g}'_e + \theta \vec{u}.\vec{g}'_e \right] dV.$$

9.3 Results from Stability Analysis

9.3.1 Plane Capacitor

The TEHD convection in a dielectric fluid between two plates in microgravity has been investigated by many authors [9–13]. It has been found that the critical modes are stationary and the corresponding critical values are $L_c = 2128.696$ and $q_c = 3.226$, where q is the wave number of the perturbations in the plane of invariance. These values, which have been confirmed by different authors, are different from the critical parameters of the Rayleigh-Bénard (RB) instability: $Ra_c = 1707.8$ and $q_c = 3.117$, where $Ra = \alpha \Delta T g d^3/\nu k$ is the Rayleigh number based on the Earth's gravity $\vec{g}$. Stiles has shown that application of electric potential to a stable configuration of fluid between two plates with an upward temperature gradient leads to an instability with a threshold L_c that increases with the value of $-Ra$ [12].

Recently, Yoshikawa et al. [8] revisited the problem of TEHD in a plane capacitor in microgravity by solving linear stability equations with consideration of the feedback effect of the temperature on the electric field. They showed that the difference in the critical parameters from the RB instability arises from stabilizing effects of the thermoelectric feedback through the perturbation electric gravity $\vec{g}'_e$; that is, W_{PG} takes a non-negligible negative value. The sensitivity of L_c and q_c to the thermoelectric parameter B has also been found:

L_c decreases as B exceeds the value of 0.3, while the critical wave number increases.

In this work, they reported that just above the threshold of TEHD convection, the heat transfer coefficient is given by $Nu = 1 + 0.78\ (L/L_c - 1)$, while for Rayleigh-Bénard convection, it is given by $Nu = 1 + 1.43(Ra/Ra_c - 1)$. This difference has been explained by the negative contribution of W_{PG} to the kinetic energy evolution: The thermoelectric feedback coupling impedes convective flow.

9.3.2 Cylindrical Capacitor

The TEHD convection in annulus has interested some researchers by the central nature of the electric buoyancy force (Table 9.1) and by its potential applications in heat transfer enhancement [3–5, 7–14]. Linear stability studies have been developed assuming the axisymmetry of convection flow. Sensitivity of the critical parameters to the direction of the temperature gradient has been found [13]. However, most of these studies assumed the small gap approximation (i.e., $\eta \sim 1$) and neglected the thermoelectric feedback.

In a recent study, Yoshikawa et al. [7] have released the small gap approximation and the assumption of axisymmetry of perturbations. They investigated the critical conditions of thermal convection for a large range of radius ratio ($0.02 < \eta < 0.999$) with the complete feedback effect. They found that the critical modes are non-axisymmetric stationary modes, although they are neither toroidal nor columnal. The critical value L_c varies significantly with η. For positive thermoelectric parameter B, the critical parameter L_c recovers the value $Ra_c = 1707.8$ of the RB instability at large η, while it converges to L_c of the plane electrode geometry (Figure 9.3). The computation of the energy generation terms W_{BG} and W_{PG} for critical modes has led to the conclusion that the basic electric gravity $\vec{g}_e$ is the driving force of the convection: The TEHD convection is analogue to the ordinary thermal convection. The thermoelectric feedback through the perturbation gravity $\vec{g}'_e$ has stabilizing effects and it becomes significant as $\eta \rightarrow 1$. The sensitivity of the critical parameters on curvature is analogous to that of thermal convection with centrifugal gravity in differentially heated annulus with solid rotation [15]. The TEHD convection in cylindrical annulus was observed in the experiment [5] for a small value of Ra and in a recent experiment on parabolic flight [16].

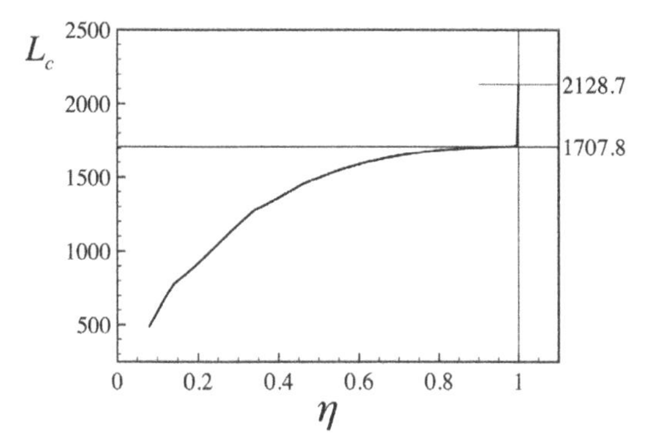

Figure 9.3 Critical electric Rayleigh number Lc for the annular geometry (B = 10^{-4}).

9.3.3 Spherical Shell

Rayleigh-Bénard convection in spherical shells with the condition of a radially directed gravitational buoyancy force is a general basis in geophysical flows, for example, for convective attributes in the inner Earth's mantle or core [17, 18]. Spherical laboratory experiments involving this configuration of a "self-gravitating" force field [19] will always be dominated by "natural" gravity, which is then vertically downward rather than radially inward. One alternative is to conduct the experiment in microgravity, thereby switching off the vertically upward buoyancy force. Supplementary, the application of the electric field as introduced above allows realizing a DEP-driven TEHD experiment as Rayleigh-Bénard analogue.

Travnikov et al. [20] perform a linear stability analysis for such a setup in microgravity environment varying the radius ratio h from 0.1 to 0.9. As a result, the eigenvalues are real, therewith delivering the independence of the Prandtl number (as non-dimensional parameter of the physical properties of the liquid). A lower curvature with $\eta \rightarrow 0.9$ leads to the higher critical onset of convection L_c and higher critical l-modes (*e.g.*, $\eta = 0.3$, $l = 2$; $\eta = 0.5$, $l = 4$; $\eta = 0.7$, $l = 7$; $\eta = 0.9$, $l = 12$). Critical values for higher η are closer nearby each other. This leads to the specific design of a microgravity experiment with $\eta = 0.5$ to avoid a critical slowing down during reaching a stable convective state. Furthermore, symmetry-breaking bifurcations by means of simulations together with path-following techniques and stability computations have been applied for $\eta = 0.5$ in [21]. The patterns of convection produce different symmetries in the form of axisymmetry, octahedral, and fivefold ones. Transition to time periodic states captures a remnant tetrahedral pattern symmetry before irregular flow appears.

In both studies [20, 21], the critical L_c is referenced to the outer radius ratio. To compare it with values from the literature, we can refer to, for example, mantle models. In a very recent work, we discuss this extensively [22]. The main objective for the comparison is the r^{-5} dependency of the electric field, which is in contrast to the geophysical models. In mantles of planetary bodies, however, the gravity is taken to be constant, whereas hydrodynamic convective modes in the Earth's liquid outer core are considered with a linear dependency. But in [21–23], we conclude that the spatial dependency is not such of relevance and that the whole scenario from the onset of convection via transition to chaos is generic.

9.4 Conclusion

The thermo-electrohydrodynamic convection in dielectric fluids represents a simple way of realizing thermal convection under microgravity conditions. The present chapter has explained the main physical mechanism underlying this convection. The key role of the electric gravity was highlighted for three simple geometries of capacitors. The critical parameters depend on the geometry parameters (curvature) and on the thermoelectric parameter. For large values of thermoelectric parameter, the perturbation gravity increases the threshold of the thermal convection. The thermo-electrohydrodynamic convection in dielectric oils has been observed in experiments and might play a growing role in the heat transfer enhancement of aerospace equipment.

Acknowledgment

The authors are grateful for the bilateral PROCOPE program. The "GeoFlow" project is funded by ESA (Grant No. AO-99-049) and by the German Aerospace Center DLR (Grant No. 50WM0122 and 50WM0822). The authors would also like to thank ESA for funding the "GeoFlow" Topical Team (Grant No. 18950/05/NL/VJ). B.F. thanks the financial support by the Brandenburg Ministry of Science, Research and Culture (MWFK) as part of the International Graduate School at Brandenburg University of Technology (BTU) and the funding at the Otto von Guericke Universität Magdeburg through the Saxony-Anhalt Ministry of Culture (MK HGB/BFP). I.M thanks the financial support from CNES (French Space Agency) and the CPER-Haute Normandie under the program THETE. H.Y and I.M. acknowledge the financial support by the french Agence Nationale de la Recherche (ANR), through the program "Investissements d'Avenir" (ANR-10 LABX-09-01), LABEX EMC^3.

References

[1] Chandrasekhar, S. *Hydrodynamic and Hydromagnetic Stability*. Dover Publications, 1981.

[2] Landau L., and E. Lifshitz. *Course of Theoretical Physics—Electrodynamics of Continuous Media,* Vol. 8. Oxford: Butterworth-Heinemann, 1984.

[3] Paschkewitz, J.S., and D.M. Pratt. "The Influence of Fluid Properties on Electrohydrodynamic Heat Transfer Enhancement in Liquids Under Viscous and Electrically Dominated Flow Conditions. *Experimental Thermal and Fluid Science* 21 (2000): 187–197.

[4] Laohalertdecha, S., P. Naphon, and S. Wongwises. "A Review of Electrohydrodynamic Enhancement of Heat Transfer." *Renewable Sustainable Energy Reviews* 11 (2007): 858–876.

[5] Chandra, B., and D. Smylie. "A Laboratory Model of Thermal Convection Under a Central Force Field." *Geophysical Fluid Dynamics* 3 (1972): 211–224.

[6] Yoshikawa, H.N., O. Crumeyrolle, and I. Mutabazi. "Dielectrophoretic Force-Driven Thermal Convection in Annular Geometry." *Physics of Fluids* 25 (2013): 024106.

[7] Yavorskaya, I.M., N.I. Fomina, and Y.N. Belyaev. "A Simulation of Central Symmetry Convection in Microgravity Conditions." *Acta Astronautica* 11 (1984): 179.

[8] Yoshikawa, H.N., M. Tadie Fogaing, O. Crumeyrolle, and I. Mutabazi. "Dielectrophoretic Rayleigh-Bénard Convection Under Microgravity Conditions." *Physical Review. E, Statistical, Nonlinear, and Soft Matter Physics* 87, no. 4 (2013): 043003.

[9] Roberts, P.H. "Electrohydrodynamic Convection." *The Quarterly Journal of Mechanics and Applied Mathematics* 22 (1969): 211–220.

[10] Turnbull, R.J., and J. R. Melcher. "Electrohydrodynamic Rayleigh-Taylor Bulk Instability." *Physics of Fluids* 12 (1969): 1160.

[11] Melcher, J.R. *Continuum Electromechanics*. Massachusetts: The MIT Press, 1981.

[12] Stiles, P.J. "Electro-Thermal Convection in Dielectric Liquids." *Chemical Physics Letters* 179 (1991): 311–315.

[13] Takashima, M. "Electrohydrodynamic Instability in a Dielectric Fluid Between Two Coaxial Cylinders." *The Quarterly Journal of Mechanics and Applied Mathematics* 33 (1980): 93–103.

[14] Stiles, P.J., and M. Kagan. "Stability of Cylindrical Couette Flow of a Radially Polarised Dielectric Liquid in a Radial Temperature Gradient." *Physica A* 197 (1993): 583–592.

[15] Busse, F.H., M.A. Zaks, and O. Brausch. "Centrifugally Driven Thermal Convection at High Prandtl Numbers." *Physica D* 184 (2003): 3–20.

[16] Dahley, N., B. Futterer, C. Egbers, O. Crumeyrolle, and I. Mutabazi. "Parabolic Flight Experiment 'Convection in a Cylinder' Convection Patterns in Varying Buoyancy Forces." *Journal of Physics: Conference Series* 318 (2011): 082003.

[17] Schubert, G., and D. Bercovici. *Treatise on Geophysics—Mantle Dynamics*. Elsevier, 2009.

[18] Schubert, G., and P. Olson, *Treatise on Geophysics—Core Dynamics*. Elsevier, 2009.

[19] Busse, F.H. "Convective Flows in Rapidly Rotating Spheres and Their Dynamo Action." *Physics of Fluids* 14 (2002): 1301–1313.

[20] Travnikov, V., C. Egbersand, R. Hollerbach. "The GEOFLOW-Experiment on ISS (Part II): Numerical Simulation." *Advances in Space Research* 32 (2003): 181–189.

[21] Feudel, F., K. Bergemann, L. Tuckerman, C. Egbers, B. Futterer, M. Gellert, and R. Hollerbach. "Convection Patterns in a Spherical Fluid Shell." *Physical Review E* 83 (2011): 046304.

[22] Futterer, B., A. Krebs, A.-C. Plesa, F. Zaussinger, D. Breuer, and C. Egbers. "Sheet-Like and Plume-Like Thermal Flow in a Spherical Convection Experiment Performed Under Microgravity." *Journal of Fluid Mechanics* 735, (2013): DOI: 10.1017/jfm.2013.507.

[23] Futterer, B., N. Dahley, S. Koch, N. Scurtu, and C. Egbers. "From Isoviscous Convective Experiment GeoFlow I to Temperature-Dependent Viscosity in GeoFlow II: Fluid Physics Experiments on-board ISS for the Capture of Convection Phenomena in Earth's Outer Core and Mantle." *Acta Astronautica* 71 (2012): 11–19.

10

The Plateau Method

Daniel A. Beysens

CEA-Grenoble and ESPCI-Paris-Tech, Paris, France

10.1 Introduction

The Plateau method has been called from the Belgium physicist Joseph Plateau. The method originates from an accidental event in 1840, when an assistant leaked some oil into a container filled with a mixture of water and alcohol. Plateau observed that the drops of oil shaped into perfect spheres in the mixture. The densities of both water–alcohol and oil phases were indeed matched by chance, thus cancelling by buoyancy thc weight of the oil drop.

10.2 Principle

The method of density matching between two immiscible phases was originally devoted to mixture of oil (the inclusion) in fully miscible water–alcohol mixtures (the host phase) [1–2]. Density of the water–alcohol host phase can be finely adjusted by varying the concentration of alcohol.

This method can be directly generalized to any method of density matching between two immiscible or partially miscible phases. Solid phases in a liquid (host) phase can also be compensated by buoyancy. The host phase has thus to be finely adjusted by different means as e.g. using completely miscible mixtures with variable concentration.

In the method, weight is compensated by buoyancy such that

$$g\triangle\rho = \varepsilon. \tag{10.1}$$

with $\triangle\rho = \rho_I - \rho_H$ the difference of density between the levitated, inclusion phase (density ρ_I) and the density-varied liquid host phase (density ρ_H). ε is a constant which has to be made as small as possible. g is the Earth's gravity acceleration constant.

In order to give an example, let us consider an investigation that needs a steady milli-g environment with two phases of density difference on order $\Delta\rho \sim 100$ kg m^{-3}. Here, $\varepsilon \sim 1$ N m^{-3} and the same environment in Earth-bound situation will correspond to a steady density matching on order $\Delta\rho \sim 0.1$ kg m^{-3}.

The host phase is made with two components 1, 2 whose mass concentration of component 1, c, is varied to have density ρ_H of the host phase adjusted. Assuming the additivity of volume, which is valid for small c, one obtains

$$\frac{1}{\rho_H} = \frac{c}{\rho_1} + \frac{1-c}{\rho_2} \tag{10.2}$$

The accurateness of the weight compensation can also be estimated by the value of the Bond number Bo. It compares the capillary to gravity forces and can be expressed by comparing the typical size R of the inclusion to the capillary length $l_c = (\frac{\sigma}{g\Delta\rho})^{1/2}$ (σ is the interfacial tension):

$$Bo = (\frac{R}{l_c})^2 = \frac{\Delta\rho R^2}{\sigma} g \tag{10.3}$$

Small values of the Bond number thus correspond to a high efficiency of the weight compensation. It is worth noting that for the same Bond number value, one can either reduce g or $\Delta\rho$ such that the product g or $\Delta\rho$ keeps constant, in a way similar to buoyancy reduction as in Equation (10.1).

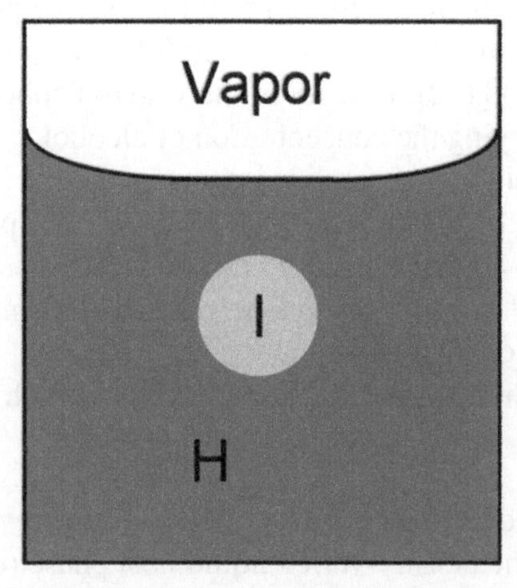

Figure 10.1 The Plateau principle. I: inclusion phase. H: host phase, made of miscible liquids whose density is adjusted by varying its concentration to match the inclusion density.

Density matching has thus to be carefully and finely adjusted by varying the concentration of a component of the host liquid. This gives a number of constraints on the method. They are concerned with temperature, aging due to the miscibility of one of the components with another (e.g., oil in alcohol), preferential evaporation of one of a component, and the general requirement of having the smallest possible variation of key parameters, for example, hydrodynamic, thermodynamic, or optical properties when varying concentration. To face these difficulties, a number of improvements have been considered. They are mainly concerned with using sets of isomeric and homologous organic compounds to match the density of pure water [3], mixtures of isotopes (cyclohexane and deuterated cyclohexane [4], water and heavy water [5]). These constraints have to be matched with other constraints specific to the experiment concerning, for example, the values of surface tension, viscosity, refractive index, etc. This method has been successfully used in the past to study the stability of liquid bridges [6], the influence of sedimentation on colloidal coagulation [5], or phase transition very near a liquid–liquid critical point [4].

10.3 Temperature Constraint

In order to achieve a density control on order of 0.1 kg m^{-3}, temperature has to be maintained constant. Typically, the temperature density variation is on order 1 kg m^{-3} K^{-1}, which means that a temperature control on order ± 0.1 °C is needed. Note that this temperature constraint gives also a further limitation on the study of phenomena where larger temperature variations are observed, for example boiling and thermocapillary motion.

In order to keep levitated the inclusion phase over a larger temperature range, one can try to also fit the temperature density variation of both inclusion and host phases. With $\alpha_i = \left(\frac{\partial \rho}{\partial T}\right)_p$ the density thermal expansion coefficient for component i = 1, 2), one has

$$\rho_i = \rho_{i0} + \alpha_i (T - T_0) \tag{10.4}$$

Here, T_0 is a reference temperature, where component density is ρ_{i0}. From Equation (10.2), one gets the thermal derivative of the host phase:

$$\alpha_H = c\left(\frac{\rho}{\rho_1}\right)^2 \alpha_1 + (1-c)\left(\frac{\rho}{\rho_2}\right)^2 \alpha_2 \tag{10.5}$$

Making $\alpha_H = \alpha_I$ necessitates a careful choice of the host components to fulfill both density matching over a large range of temperature. An example is known in a partially miscible liquid mixture of methanol and partially deuterated cyclohexane. The levitation of the methanol (inclusion) phase is ensured over a large range of temperature in a cyclohexane +5.5% partially deuterated cyclohexane mixture (host phase) as α_H = –0.96 kg m^{-3} K^{-1} and α_I = –0.95 kg.m^{-3} K^{-1}.

10.4 Other Constraints

In the classical Plateau method where three components are present (oil, water, alcohol), there is a significant change of density matching with time as alcohol either evaporates or mixes with the oil phase. To overcome this difficulty, a number of improvements have been proposed:

Using only two immiscible liquids of equal density. A large number of transparent organic liquid pairs immiscible with water have been listed with their physical data in Refs. [3, 7]. Such compounds are numerous, but most show some miscibility with water (e.g., dibutyl phthalate or isobutyl benzoate). However, with only two components, fine density adjustment can be made only by temperature changes to improve the tuning.

Using sets of isomeric and homologous organic compounds. The compositional change with time can be diminished by using sets of isomeric and homologous organic compounds to match the density of pure water [3].

10.5 Concluding Remarks

The Plateau method, based on density matching, is appealing by its simplicity for compensating buoyancy on Earth. It is, however, limited to mixture of immiscible liquids and to solids in liquids. A number of difficulties have to be overcome to obtain reliable data, for example, the difference of temperature variation of host and inclusion phase densities, weak miscibility between phases, and differential evaporation of phases. These difficulties can in general be solved, which makes the Plateau method a very useful and largely employed technique for compensating gravity on Earth.

References

[1] Plateau, J., "Experimental and Theoretical Researches on the Figures of Equilibrium of a Liquid Mass Withdrawn from the Action of Gravity." *Annual Report of the Board of Regents of the Smithsonian Institution* (1863): 207–285.

[2] Plateau, J., *Statique Expérimentale et Théorique des Liquides Soumis aux Seules Forces Moléculaires*. Paris: Gauthier-Villars, 1873.

[3] Lowry, B.J., and P. H. Steen, "On the Density Matching of Liquids Using Plateau's Method." *Proceedings of the 32nd Aerospace Sciences Meeting & Exhibit*, AIAA 94-0832, 1994; DOI: 10.2514/6.1994–832

[4] Houessou, C., P. Guenoun, R. Gastaud, F. Perrot, and D. Beysens. "Critical Behavior of the Binary Fluids Cyclohexane-Methanol, Deuterated Cyclohexane-Methanol and of Their Isodensity Mixture. Application to Microgravity Simulations and Wetting Phenomena." *Physical Review A* 32 (1985): 1818–1833.

[5] Liu, Ji., Zhi Wei Sun, and AA Yan. "Non Gravitational Effects with Density Matching in Evaluating the Influence of Sedimentation on Colloidal Coagulation." *Chinese Physics Letters* 22 (2005): 3199–3202.

[6] Lowry, B.J., and P. H. Steen. "Stability of Slender Liquid Bridges Subjected to Axial Flows." *Journal of Fluid Mechanics* 330 (1997): 189–213.

[7] Smedley, G., and D. Coles. "Some Transparent Immiscible Liquid Pairs." *Journal of Colloid and Interface Science* 138 (1990): 42–60.

11

Centrifuges

Jack J. W. A. van Loon

VU Medical Center, Amsterdam, The Netherlands

11.1 Introduction

Research in orbital space flight is using to a large extent the fact that gravity is compensated by the spacecraft's free fall around the globe. Reducing gravity to a minimum (weightlessness) has great advantages with respect to the decrease in mechanical forces, decreased convection, buoyancy, hydrostatic pressure, etc. Minimizing these phenomena also provides opportunities to study processes otherwise obscured at terrestrial gravity acceleration levels ($1g$) such as Marangoni convection or studies involving materials phase shift phenomena.

The use of centrifuges in this field of research is motivated by interest in the impact of weight on systems. One thus needs to explore the full range of the gravity spectrum, from the theoretical zero g up until a certain maximum, whatever that maximum is for the system under study.

When we consider physical parameters acting upon a system, the factor weight is basically not any different from, for example, temperature or pressure, and in order to understand how the systems respond to environmental variables, we need to modulate them. For many systems, it is therefore as relevant to look at hypergravity (any value above Earth $1g$) as well as hypogravity (between 1 and μg) or even near weightlessness.

Gravity generated in a centrifuge is caused by inertia where an object in motion will move at the same speed in the same direction unless forced to change its direction. This is exactly what a centrifuge does; it forces the object to constantly change its direction. This causes acceleration or when compared to acceleration brought about by, for example, the mass of Earth, artificial gravity.

In a constantly rotating centrifuge, the object moves with a constant velocity. However, since the orientation is constantly changed, the object is submitted to the centripetal acceleration $a_c = \omega^2 r$, where ω is the angular velocity (rad s^{-1}) and r the centrifuge arm radius. An object thus experiences the centripetal force $F_c = ma_c = m\omega^2 r$.

Centrifuges/hypergravity is used in both life and physical sciences. Surely, in the process industry and laboratories, centrifuges are widely applied to separate substances from different specific densities being it particles in liquids, liquids with different specific densities [1] or radioisotopes [2]. Sometimes, centrifuges are even helpful to shed light on geological historical events with biblical significance [3]. In diamond synthesis research, both microgravity [4] and hypergravity [5] are used and both environments have their specific benefits.

Over the years, operating the large-diameter centrifuge (LDC) (Figure 11.1 right [6]) at ESA-ESTEC (Noordwijk, The Netherlands), we have performed experiments in both life and physical sciences' domains. These range from, for example, cell biology [7], plant [8] and animal physiology [9, 10] to granular matter [11], geology/planetary sciences [12], and fluid [13] or plasma physics [14]. In all these studies, the impact of weight was explored from 1 up to $20g$. Doing so, one could also try to explore the influence of gravity at lower g levels by extrapolating the hypergravity data. Such extrapolation has been proposed earlier via the so-called gravity

Figure 11.1 Two examples of research centrifuges. *Left* The medium-diameter centrifuge for artificial gravity research (MidiCAR) is a 40-cm-radius system for (mainly) cell biology research [16]. *Right* The large-diameter centrifuge (LDC) [6] is an 8-meter-diameter system used for life and physical sciences and technological studies. Both centrifuges are located at the TEC-MMG Lab at ESA-ESTEC, Noordwijk, The Netherlands.

continuum [15]. See for a great publication on artificial gravity in human research the book by Cl6ment and Bukley [16].

11.2 Artifacts

11.2.1 Coriolis

One of the side effects, or artifacts, of rotating systems, like gravity generated in a centrifuge, is the Coriolis effect. Some call it the Coriolis force, but actually it is not a force.

These Coriolis accelerations are experienced by objects which move within a rotating system. It is an effect of rotation that contributes to the impurity of gravity generated by the rotating system.

$$a_{\mathrm{Corliolis}} = (2\upsilon \times \omega)g$$

where a_{Coriolis} is the Coriolis acceleration expressed in units of g, υ = velocity of the moving object, and ω = angular velocity of the rotating system. When the angular velocity is expressed in revolutions per minute (rpm),

$$a_{\mathrm{Corliolis}} = (2\pi\upsilon \times \omega)g.$$

The extend of the Coriolis effect on a sample depends on the axes along which the object moves with respect to the rotation axes. The largest impact is when the motion of the sample occurs in a plane at a 90° angle with relation to the rotation axes. The impact of the effect is relatively larger in fast rotating systems. Therefore, one likes to work with large-diameter centrifuges in order to keep this variable as small as possible. Two rotating systems with different radii but spinning at similar speeds would generate the same Coriolis accelerations on an object that moved with the same speed within such rotating systems. For more details regarding life sciences, see, for example, van Loon [18], and for physical sciences, see, for example, Battaile et al. [19].

Particularly, in vestibular research, the phenomena of cross-coupled angular acceleration is apparent. This occurs while a person rotates the head in another plain than that of the rotating device. This results in an apparent rotation in a direction unrelated to what is actually happening and is perceived as tumbling, rolling, or yawing. The phenomenon results from simultaneous rotation about two perpendicular axes, and the magnitude of this type of acceleration is the product of the two. Cross-coupled acceleration is one of the main courses of motion sickness in rotating systems. See also Antonutto et al. [20] and Elias et al. [21].

11.2.2 Inertial Shear Force

Another quite common artifact in rotating systems is inertial shear. This effect is mainly experienced by samples that are attached to a flat surface fixed in a centrifuge. Since a centrifuge, clinostat, or RPM is describing curved or round trajectories, also the surfaces used, for example to culture cells, need to be shaped to the same curvature as the rotating device. This means cells on a flat surface in a large-radius rotating system experience less inertial shear as the same cells on the same surface in a smaller-diameter system both generating the same *g*-level. The laterally directed inertial shear confounds the actual centrifugal force which acts perpendicular to a sample. This phenomenon and artifact was identified in 2003 [22]. Although we predicted an impact of this artifact for biological systems, no in situ evidence had been shown until recently. It became clear that using large flat surfaces in a fast rotating system (a clinostat) generated inertial shear forces within the cell layers studied [23].

For completeness, we need to mention that inertial shear force is, however, not restricted to rotating system but should also be taken into account with fast accelerating or decelerating linear systems or in vibration studies.

11.2.3 Gravity Gradient

Besides the inertial shear due to lateral forces, one also has to take into account an artifact in the axial direction: the gravity gradient. A volume exposed to gravity within a centrifuge experiences more acceleration further from the center of rotation. Also here, the larger the centrifuge, the smaller the impact of this artifact. Centrifuges are not only used on-ground but also in-flight. In-flight systems are either used to generate an in-flight $1g$ control in order to discriminate between, for example, launch and operations effects and cosmic radiation [24]. In a facility such as the European Modular Cultivation System (EMCS) tailored for plant studied, full-grown plants will have a huge *g*-gradient of +40 and –40 % around the $1g$ center due to its limited radius [24]. This also applies for the application of small-radii human centrifuges that are currently explored as a possible in-flight countermeasure [25]. The gravity gradient in such centrifuges can be as much as over 300 %. Again, large radii would be the credo. A novel and challenging project explores the possible applications for a large ground-based human hypergravity habitat (H^3) that focuses on a centrifuge diameter of some 150–200 m. In this H^3 facility, where humans can live in for periods of weeks or month at constant hypergravity, both the Coriolis and body *g*-gradient are reduced to a minimum [26, 27].

Figure 11.2 View of the outside structure that accommodates the envisaged human hypergravity habitat (H^3). The H^3 is a large-diameter (~175 m) ground-based centrifuge where subjects can be exposed to higher g-levels for periods up to weeks or months. The H^3 can be used in preparation for future human exploration programs as well as for regular human physiology research and applications [26, 27]. (*Image* © BERTE bvba/van Loon et al. 2012).

11.3 The Reduced Gravity Paradigm (RGP)

Particularly, in life sciences cells, tissues or whole organisms respond to gravity based on, most likely, their mass and mechanical properties by changes in, for example, morphology. We see this in cells which change their growth, morphology, and extracellular matrix when cultured at high g loads (e.g. [28]). Also, the cytoskeleton protein actin is increased at increased g-level [29], while actin fiber levels are decreased in real microgravity [30]. In human being, we see an increase in vertebral disk height with decreased gravity loads [31], while disk height decreases with an increased mechanical load [32]. But also in plants, we see an increase in the structural component lignin at higher g-levels compared to $1g$ controls [33].

Based on these observations, I would like to introduce the "reduced gravity paradigm (RGP)". The paradigm would provide insight on the changes seen while going into space (from $1g$ to μg) based on the observations while going from hyper-g to $1g$ condition.

The premise for this paradigm is that the system under study should have reached some sort of steady state at a higher g-level before transferring to a lower g-level. In practice, this would imply starting the experiment in a centrifuge and after some time stop centrifugation, or reducing the centrifugal force to some level. The sample will adapt to the reduced g force. The presumption is that the adaptations during this period are the same, or at least similar, to changes seen while going into real microgravity coming from *terra firma*. Basically, we use centrifuges for microgravity simulation.

A first indication for this paradigm was from a very simple cell biological experiment we did in 2002 where we exposed bone cells to high g forces in the MidiCAR centrifuge [34]. After some time, the centrifuge was stopped, and the cells retrieved and fixed. They were analyzed for cell height. From these experiments, we initially concluded that cell heights increased with increasing g-levels. However, we expected cell heights to decrease with increasing g. To verify this hypothesis, we also conducted a more sophisticated experiment where we measured cell height by means of an atomic force microscope (AFM) accommodated inside a centrifuge. Cell heights were measured while going through three g-levels, 1, 2, and $3g$ [35]. The AFM data showed that the cells were indeed flattened at higher g-levels, verifying our initial hypothesis. However, this did not explain the initial increase in cell heights from the first experiments. What might have happened is that the cells in the initial study were indeed flattened at high g but displayed a kind of overshoot recovery after stopping the centrifuge and before fixation. This response to a reduced gravity might indicate what might happen when cells are exposed to real microgravity.

Such experimental designs could be applied in all fields of research and especially in life sciences where cells and tissues respond, in general, in a more gradual manner than, for example, in purely physical sciences phenomena. Prerequisite for such a "microgravity simulation"—"reduced gravity paradigm" experiment would be that one starts to enter the reduced gravity phase coming from a steady state at a higher g-level. The applicability of this RGP is especially useful for relatively rapid processes. In such processes, we can study the response to a lower g load, while on slower processes, the signals are fainted by the present $1g$ environment. Many, if not all, hypergravity studies are terminated by quickly stopping the centrifuge after which samples are fixed and processed. One might consider to add some additional groups to hyper-g studies and carefully study the readaptation process after the hypergravity loads. The paradigm was introduced at the COSPAR meeting in 2010 [36].

Besides the indication of the RGP by the cell biological experiment mentioned earlier, there has also been a very nice indication of this paradigm from a human physiology study. Quite some astronauts experience motion sickness symptoms referred to as space adaptation syndrome (SAS). Spacelab-D1 crew members together with researchers from TNO (Soesterberg, The Netherlands) were able to "reproduce" these specific in-flight phenomena by exposing astronauts to a 1.5-h +3*g* supine centrifuge run which evoked a sickness induced by centrifugation (SIC) [37, 38]. It appeared that this SIC protocol very closely resembles the in-flight experience by the crew members. So also here the immediate post-centrifuge period displays phenomena similar or even the same as seen in real microgravity, supporting this reduced gravity paradigm.

This RGP might also shed some light on experiments conducted in parabolic flights where we have a series of cyclings from 1*g* to hyper-*g*, to micro-*g* to hyper-*g* and back to 1*g* again. The data gathered in the various gravity stages might be "contaminated" by the previous *g* history. So measuring an effect in micro-*g* might be the actual effect at micro-*g*, but it there might also include some contribution from the transition from $\sim$1.8*g* to a lower *g* period.

There is quite some debate on the reliability of ground-based facilities to simulate a real microgravity environment. Discussions are focused on the one hand on fluid motions and shear stress within clinostats and RPMs and rotating wall vessels (e.g., [39, 40]) or on the other hand on the impact of the magnetic field in levitation studies (e.g., [41]). The advantage of the reduced gravity paradigm is that such artifacts are not present in such studies, other than possible post-rotation effects. The first indication of the applicability of the RGP or "relative microgravity" in a biological study is shown in a zebra fish study by Aceto et al. [42].

References

[1] Parker, P.M. *The 2013–2018 World Outlook for Laboratory Centrifuges*. France: ICON Group International Inc., 2013.

[2] Malonet, J.O. "Centrifugation". *Industrial and Engineering Chemistry* (1948).

[3] Haigh, S.K. and S.P.G. Madabhushi. "Dynamic Centrifuge Modelling of the Destruction of Sodom and Gomorrah". In Proceedings of International Conference on Physical Modelling in Geotechnics. Newfoundland, Canada: St John's, July, 2002.

[4] Ishikawa, M., S. Kamei, Y. Sato, N. Fujimori, N. Koshikawa, K. Murakami and T. Suzuki. "A Diagnostic Study of Plasma CDV Under

Microgravity and its Application to Diamond Deposition". *Advances in Space Research* 24 (1999): 1219–1223.

[5] Abe, Y., G. Maizza, S. Bellingeri, M. Ishizuka, Y. Nagasaka and T. Suzuki. "Diamond Synthesis by High-Gravity dc Plasma CDV (HGCDV) with Active Control of the Substrate Temperature". *Acta Astronautica* 48 (2001): 121–127.

[6] van Loon, J.J.W.A., J. Krause, H. Cunha, J. Goncalves, H. Almeida and P. Schiller. "The Large Diameter Centrifuge, LDC, for Life and Physical Sciences and Technology". In *Proceedings of the 'Life in Space for Life on Earth Symposium'*, Angers, France, 22–27 June 2008. ESA SP-663.

[7] Ciofani, G., L. Ricotti, J. Rigosa, A. Menciassi, V. Mattoli and M. Monici. "Hypergravity Effects on Myoblast Proliferation and Differentiation". *Journal of Bioscience and Bioengineering* 113, no. 2 (2012): 258–261.

[8] Chebli, Y., L. Pujol, A. Shojaeifard, L. Brouwer, J.J. van Loon and A. Geitmann. "Cell Wall Assembly and Intracellular Trafficking in Plant Cells are Directly Affected by Changes in the Magnitude of Gravitational Acceleration". *PLoS One* 8, no. 3 (2013): e58246.

[9] Fraser, P.J., D.K. Reynolds, F.E. O'Callaghan and J.J.W.A. van Loon. "Effects of Hypergravity up to 20g on Eye Motor Neurones of the Crab *Carcinus maenas* (L.) During Oscillation Around the Horizontal Axis Using the ESA Large Diameter Centrifuge". *Journal of Gravitational Physiology* 17, no. 1 (2010): P1–P2.

[10] Serrano, P., J.J.W.A. van Loon, F.J. Medina and R. Herranz. "Relation Between Motility Accelerated Aging and Gene Expression in Selected *Drosophila* Strains Under Hypergravity Conditions". *Microgravity Science and Technology* 25 (2013): 67–72.

[11] Dorbolo, S., L. Maquet, M. Brandenbourger, F. Ludewig, G. Lumay, H. Caps, N. Vandewalle, S. Rondia, M. Melard, J.J. van Loon, A. Dowson and S. Vincent-Bonnieu. "Influence of the Gravity on the Discharge of a Silo". *Granular Matte*r 15, (2013): 263–273.

[12] Gibbings, A., E. Komninou and M. Vasile. "Investigation and Modeling of Large Scale Cratering Evens—Lessons Learnt from Experimental Analysis". AIAC-11.E1.8.5. 62nd IAC, Cape Town, South Africa, 2011.

[13] Lioumbas, J.S., J. Krause and T.D. Karapantsios. "Hypergravity to Explore the Role of Buoyancy in Boiling in Porous Media". *Microgravity Science and Technology* 25, no. 1 (2013): 17–25.

[14] Sperka, J., P. Soucek, J.J.W.A. van Loon, A. Dowson, C. Schwarz, J. Krause, G. Kroesen and V. Kudrle. "Hypergravity Effects on

Glide Arc Plasma". *The European Physical Journal* D 67 (2013): 261.

[15] Plaut, K., R.L. Maple, C.E. Wade, L.E. Baer, A.E. Ronca. "Effects of Hypergravity on Mammary Metabolic Function: Gravity Acts as a Continuum". *Journal of Applied Physiology* 95 (2003): 2350–2354.

[16] Clément, G. and A. Bukley, eds. *Artifical Gravity*. Springer, 2006.

[17] van Loon, J.J.W.A., L.C. van den Bergh, R. Schelling, J.P. Veldhuijzen and R.H. Huijser. "Development of a Centrifuge for Acceleration Research in Cell and Developmental Biology". In *44th International Astronautical Congress*, IAF/IAA-93-G.4.166, Gratz, Austria, October 1993.

[18] van Loon, J.J.W.A. Chapter-1 "The gravity environment in space experiments". In *Biology in Space and Life on Earth*, edited by E. Brinckmann. Weinheim, Germany: Whiley,. 2007, pp. 17–32.

[19] Battaile, C.C., R.N. Grugel, A.B. Hmelo and T.G. Wang. "The Effect of Enhanced Gravity Levels on Microstructural Development in Pb50 wt pct Sn Alloys During Controlled Directional Solidification". *Metallurgical and Materials Transactions A* 25, no. 4 (1994): 865–870.

[20] Antonutto, G., D. Linnarsson, C.J. Sundberg and P.E. di Prampero. "Artificial Gravity in Space Vestibular Tolerance Assessed by Human Centrifuge Spinning on Earth". *Acta Astronautica* 27 (1992): 71–72.

[21] Elias, P., T. Jarchow and L.R. Young. "Modeling Sensory Conflict and Motion Sickness in Artificial Gravity". *Acta Astronautica* 62 (2007): 224–231.

[22] van Loon Jack, J.W.A., Erik H.T.E. Folgering, Carlijn V.C. Bouten, J. Paul Veldhuijzen and Theo H. Smit. "Inertial Shear Forces and the Use of Centrifuges in Gravity Research. What is the Proper Control?" *ASME Journal of Biomechanical Engineering* 125, no. 3 (2003): 342–346.

[23] Warnke, E., J. Pietsch, M. Wehland, J. Bauer, M. Infanger, M. Görög, R. Hemmersbach, M. Braun, X. Ma, J. Sahana and D. Grimm. "Spheroid Formation of Human Thyroid Cancer Cells Under Simulated Microgravity: a Possible Role of CTGF and CAV1". *Cell Commun Signal* 12 (2014): 32.

[24] van Loon, J.J.W.A., E.H.T.E. Folgering, C.V.C. Bouten and T.H. Smit. "Centrifuges and Inertial Shear Forces". *Journal of Gravitational Physics*. 11, no. 1 (2004): 29–38.

[25] Iwase, S., J. Sugenoya, N. Nishimura, W.H. Paloski, L.R. Young, J.J.W.A. van Loon, F. Wuyts, G. Clément, J. Rittweger, R. Gerzer and

J. Lackner. "Artificial Gravity with Ergometric Exercise on International Space Station as the Countermeasure for Space Deconditioning in Humans". In *31st Annual ISGP Symposium, 11th ESA Life Sciences Symposium, 5th ISSBB Meeting, ELGRA Symposium*, Trieste, Italy, June 2010.

[26] van Loon, J.J.W.A. "The Human Centrifuge". *Microgravity Science and Technology* 21 (2009): 203–207.

[27] van Loon, J.J.W.A., J.P. Baeyens, J. Berte, S. Blanc, L. Braak, K. Bok, J. Bos, R. Boyle, N. Bravenboer, E.M.W. Eekhoff, A. Chouker, G. Clement, P. Cras, E. Cross, M.-A. Custaud, M. De Angelis, T. Delavaux, R. Delfos, C. Poelma, P. Denise, D. Felsenberg, D. Delavy, K. Fong, C. Fuller, S. Grillner, E. Groen, J. Harlaar, M. Heer, N. Heglund, H. Hinghofer-Szalkay, N. Goswami, M. Hughes-Fulford, S. Iwase, J.M. Karemaker, B. Langdahl, D. Linnarsson, C. Lüthen, M. Monici, E. Mulder, M. Narici, P. Norsk, W. Paloski, G.K. Prisk, M. Rutten, P. Singer, D.F. Stegeman, A. Stephan, G.J.M. Stienen, P. Suedfeld, P. Tesch, O. Ullrich, R. van den Berg, P. Van de Heyning, A. Delahaye, J. Veyt, L. Vico, E. Woodward, L.R. Young and F. Wuyts. "A Large Human Centrifuge for Exploration and Exploitation Research". *Annales Kinesiologiae* 3, no. 1 (2012): 107–121.

[28] Croute, F., Y. Gaubin, B. Pianezzi and J.P. Soleilhavoup. "Effects of Hypergravity on the Morphology, the Cytoskeleton, the Synthesis of Extracellular Macromolecules and the Activity of Degradative Enzymes". *Life Sciences in Space Research*. ESA SP-366 (1994): 31–24.

[29] Li, S., Q. Shi, G. Liu, W. Zhang, Z. Wang, Y. Wang and K. Dai. "Mechanism of Platelet Functional Changes and Effects of Anti-Platelet Agents on In Vivo Hemostasis Under Different Gravity Conditions". *Journal of Applied Physiology* 108, no. 5(2010): 1241–1249.

[30] Vassy, J., S. Portet, M. Beil, G. Millot, F. Fauvel-Lafeve, A. Karniguian, G. Gasset, T. Irinopoulou, F. Calvo, J.P. Rigaut and D.S. "The Effect of Weightlessness on Cytoskeleton Architecture and Proliferation of Human Breast Cancer Cell Line MCF-7". *FASEB J* 15 (2001): 1104–1106.

[31] Macias, B.R., P. Cao, D.E. Watenpaugh and A.R. Hargens. "LBNP Treadmill Exercise Maintains Spine Function and Muscle Strength in Identical Twins During 28-Day Simulated Microgravity". *Journal of Applied Physiology* 102, no. 6(1985): 2274–2278.

[32] Hämäläinen, O., H. Vanharanta, M. Hupli, M. Karhu, P. Kuronen and H. Kinnunen. "Spinal Shrinkage due to +Gz Forces". *Aviation, Space, and Environmental Medicine* 67, no. 7(1996): 659–661.

[33] Wakabayashi, K., K. Soga, S. Kamisaka and T. Hoso. "Increase in the Level of Arabinoxylan–Hydroxycinnamate Network in Cell Walls of Wheat Coleoptiles Grown Under Continuous Hypergravity Conditions". *Physiologia Plantarum* 125, no. (2005): 127–134.

[34] van Laar Meie, C. "Influence of Gravity on Cells". Master thesis, VU-University Amsterdam, August 2002.

[35] van Loon, J.J.W.A., M.C. Van Laar, J.P. Korterik, F.B. Segerink, R.J. Wubbels, H.A.A. De Jong and N.F. Van Hulst. "An Atomic Force Microscope Operating at Hypergravity for In Situ Measurement of Cellular Mechano-response". *Journal of Microscopy* 233, no. 2 (2009): 234–243.

[36] Van Loon, J. "The Application of Centrifuges in "Reduced Gravity" Research". In *Session F12-0018-10 38th Assembly of the Committee on Space Research (COSPAR)*, Bremen, Germany, July 2010.

[37] Ockels, W.J., R. Furrer and E. Messerschmid. "Simulation of Space-Adaptation Syndrome on Earth". *ESA Journal* 13, no. 3 (1989): 235–239.

[38] Bles, W., B. de Graaf, J.E. Bos, E. Groen and J.R. Krol. "A Sustained Hyper-g Load as a Tool to Simulate Space Sickness". *Journal of Gravitational Physiology* 4, no. 2 (1997): P1–P4.

[39] Leguy, C.A.D., R. Delfos, M.J.B.M. Pourquie, C. Poelma, J. Krooneman, J. Westerweel and J.J.W.A. van Loon. "Fluid Motion for Microgravity Simulations in a Random Positioning Machine". *Gravitational and Space Biology* 25, no. 1 (2011): 36–39.

[40] Olson, W.M., D.J. Wiens, T.L. Gaul, M. Rodriguez and C.L. Hauptmeier. "Xenopus Development from Late Gastrulation to Feeding Tadpole in Simulated Microgravity". *The International Journal of Developmental Biology* 54, no. 1 (2010): 167–174.

[41] Moes, M.J.A., J.C. Gielen, R.-J. Bleichrodt, J.J.W.A. van Loon, P.C.M. Christianen and J. Boonstra. "Simulation of Microgravity by Magnetic Levitation and Random Positioning: Effect on Human A431 Cell Morphology". *Microgravity Science and Technology* 23, no. 2 (2011): 249–261.

[42] Aceto, J., R. Nourizadeh-Lillabadi, M. Marée, N. Jeanray, L. Wehenkel, P. Aleström, J.J.W.A. van Loon and M. Muller. "Gravitational Effects on Zebrafish Bone and General Physiology are Revealed by Hypergravity Studies". 2014 (submitted).

3

Facilities to Mimic Micro-Gravity Effects

12

Animals: Unloading, Casting

Vasily Gnyubkin and Laurence Vico

INSERM U1059, LBTO, Faculty of Medicine, University of Lyon,
Saint-Etienne, France

12.1 Introduction

The aim of this section is to provide description of techniques of on-ground microgravity simulation based on animal models such as hindlimb unloading, casting, and denervation.

It is well known that exposure to microgravity leads to notable restrictions in general movement and mechanical loading in astronauts. Conditions of spaceflight together with spacecraft environment, confinement, altered diet and altered ambient atmosphere, and relatively high radiation result in significant alterations in normal physiological processes. Existing countermeasures, based on physical exercises, are not able to completely substitute normal Earth gravity loading. It is undoubtedly true that the development of new countermeasures is a crucial step on a way to the long-term space missions.

One of the problems with spaceflight experiments is that opportunities to carry them out are expensive and rare. That is where animal ground-based models come into play. With ground-based models, there are no limitations related to number of animals. What is also important is that there is no ideal imitation of all conditions of long-term spaceflight. Even techniques such as head-down-tilt bed-rest studies and water immersion which are generally accepted as the gold standard imitate only some of the spaceflight conditions [see also Chapter 13]. At the same time, parabolic flights and drop towers provide weightlessness, but only for very short periods of time [see also Chapter 6 and Chapter 7].

In the past, varieties of mammalian species, including monkeys, dogs, and rabbits, were used for research purposes. Nowadays, rodents have become one of the most used animals in all areas of scientific studies. There are many reasons in favor of using them instead of primates or rabbits: mice and rats grow fast and reproduce quickly, and it is easy to house and maintain them. Also, with rats and especially mice, it is possible to conduct uniform studies with genetically identical animals. As it is easier for the mice to be genetically modified, it is also easier to breed either transgenic or knockout animals.

In order to develop an acceptable ground-based model for the simulation and study of spaceflight aspects, NASA-Ames Research Center has formulated the following requirements: experimental animals should demonstrate physiological response similar to that during spaceflight; the model should provide thoraco-cephalic fluid shift; the model should unload limbs without motion restriction or paralysis, and provide ability to recover; and the model should not be stressful for animals. Such technique would be valuable for predicting the effects of spaceflight, studying possible mechanisms of these effects, and developing countermeasures [1].

Nowadays, different immobilization techniques are widely used for the simulation of mechanical unloading. Immobilization itself can be combined with dietary or pharmaceutical intervention. Generally, methods can be merged into two groups: conservative (bandaging, casting, hindlimb unloading, and confinement) and surgical (nerve resection, denervation with botulin toxin, spinal cord resection, and tendon resection). Immobilization provided by casting, denervation, and tendon resection is widely used for the quick development of disuse osteopenia or muscle atrophy. Therefore, these models are useful for studying different countermeasures against bone or muscle loss. However, this approach has serious limitations: surgical intervention or casting does not mimic effects of spaceflight on cardiovascular system, nervous system, and immune system. In addition, with the existing surgical models, recovery from disuse is impossible or difficult. Such surgical models may also result in inflammation, altered trophic, perfusion, and innervation of immobilized limb [2].

Among microgravity simulation models, hindlimb unloading fits most of the NASA requirements. It induces muscle atrophy and alterations in bone structure similar to physiological consequences observed in humans after spaceflight or bed rest. Other physiological changes similar to spaceflight such as synaptic plasticity changes [3] and immune system suppression have also been reported in this model [4]. In cardiovascular functions, rodent head-down-tilt simulates cephalic fluid redistribution and hypovolemia. It also leads to vessel's structural and functional adaptations and alters baroreflex function [5]. Putting all these factors together, it is clear why the use of tail traction in the hindlimb unloading model has become the technique of choice for studying spaceflight-like changes in rats and mice.

12.2 Hindlimb Unloading Methodology

Emily R. Morey-Holton has done significant work on the development and standardization of hindlimb unloading. The review of technical aspects of the method produced in 2002 has been widely used as a base for microgravity simulation studies [2]. Here we provide the description of the method based on this review. Before the experiment, animals are acclimated to their cages for at least two days prior to the hindlimb unloading. First, a strip of traction tape, pre-attached to the plastic tab, is attached to the pre-cleaned tail just above the hair line. Then, the traction tape is fixed by two strips of filament tape placed around the base of the tail and on about half-way up the traction tape. To protect the traction tape, gauze bandage can be wrapped around the tail. The gauze bandage should not cover the whole tail, because the tail plays an important role in thermoregulation. Daily health checks confirmed that the exposed tip of the tail remained pink, indicating adequate blood flow [6]. The animal is then attached to the top of the cage. Such way of harnessing aims to distribute the load along the length of the tail and avoid excessive tension on a small area.

The body of the animal makes about a 30° angle with the floor of a cage, and thus the animal does not touch the grid floor with its back feet (Figure 12.1).

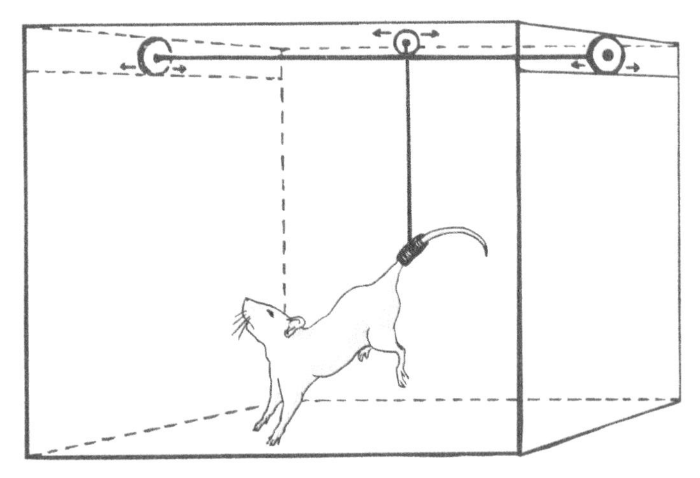

Figure 12.1 Hindlimb unloading model.

At this position, 50% of rat's body weight applies to its forelimbs. A 30° angle of unloading is recommended because it provides normal weight bearing on the forelimbs, unloads the lumbar vertebrae but not the cervical vertebrae, and induces a cephalic fluid shift [2]. The angle and height of the animals are checked, and then adjusted if necessary, on a daily basis. In order to use the system on animals with different behavioral pattern and smaller size of the body, such as mice, some adjustments to the hindlimb unloading system are needed. These include the use of smaller cages, and inclusion of a device to prevent mice from climbing the harness and chewing it. As the mice are generally weaker than rats and have smaller body weight in relation to the total weight of the unloading system, they find it difficult to freely move around the cage and have access to food and water. Therefore, a significant adjustment to the friction is needed between the roller and the wire.

There are some disadvantages of the method: tail examination can be difficult due to the size of gauze bandage; animals can chew the traction tape and release themselves; and tail can undergo inflammation or necrosis [7]. Therefore, in the literature, several modified techniques are suggested to harness fixation based on minimally invasive surgery. As one of the methods, the authors have made a harness for hindlimb unloading by inserting a surgical steel wire through intervertebral disc space of the tail. After that, the wire was ring shaped and used for suspension. Detailed step-by-step video description is publicly available on the Internet [7].

Similar method was proposed for long-term studies on adult rats. Continuous hindlimb unloading for longer than 3 weeks can be complicated in rats with high body weight (350 g or more). Animals come down from suspension because of sloughing of tail skin. Even very short periods of reloading induce changes in muscle physiology. Therefore, frequent release of animals from suspension apparatus can compromise a study [8]. In such case, after passing a steel cable through rat's tail skin and wrapping it loosely with gauze, authors used a 5-ml syringe cut in half longitudinally together with orthopedic casting to provide strong structural integrity. This was performed to ensure that the tail remained in a straight line with respect to the body once the animal was hindlimb suspended [8].

These methods are principally close to the "classical" tail suspension. Interestingly, there is a new method that uses a different unloading model. Partial weight suspension was described in Erica B. Wagner's paper in 2010 [9]. The main difference is that this system allows the distribution of gravitation loading among hindlimbs and forelimbs in a desired proportion between 10 and 80% of total weight bearing, with an accuracy of ±5%.

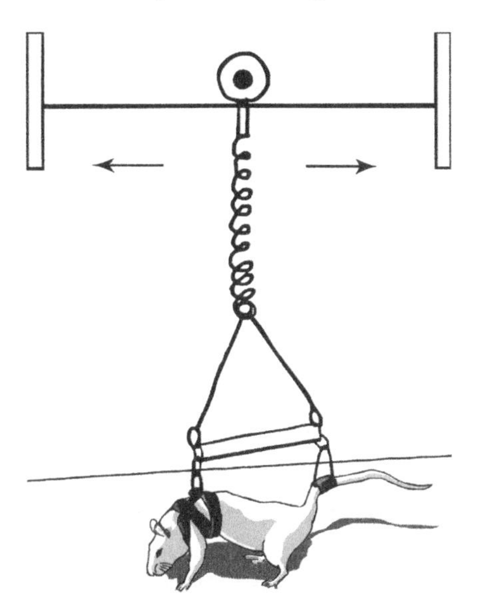

Figure 12.2 Partial weight suspension model.

With such a technique, animals have linear freedom of motions; feeding and cleaning are easy; and animals can be exposed to quadrupedal unloading for at least three weeks Figure 12.2. Also, this system allows for a full recovery of animals after an experiment [9].

To fix an animal in the desired position, two harnesses are used. One is a ring of bandage put around the tail's base. The second one is a flexible, breathable "jacket" secured around the chest cage. The tail's and the chest's harnesses are connected to adjustable chains and hollow metal rod. With this model, researchers performed a fascinating study where they imitated gravitational conditions of Mars planet [9]. Possible disadvantages of this model of microgravity simulation include an absence of head-down-tilt and physiological changes related to it.

12.3 Recommendations for Conducting Hindlimb Unloading Study

While conducting the research on animal models, it is important to remember that any interference of normal life activity is a stress for animals. Hindlimb unloading, restraint stress, and social isolation cause significant perturbations in blood pressure, heart rate levels, [5] and plasma corticosterone level [10].

Therefore, it is recommended to use minimal restraints and avoid unnecessary manipulations during preparatory period and period of tail suspension. Physiological and environmental parameters, including body weight, room temperature, and angle of unloading, should be monitored on a daily basis [11].

Animals from a control group should be kept in identical cages. Behavioral or physiological modifications produced by environmental variables can cause false results or give wrong hypothesis [11]. Another important factor related to control group is feeding. Unloaded animals lose weight during the experiment despite easy access to food and water because of the alterations in energy balance. The difference in weight between experimental and control groups, both fed ad libitum, can be from 5 to 20% in adult rats [2]. Hence, it is recommended to either feed control group with average amount of food consumed by suspended group or reduce caloric intake for control group. However, the latter can result in physiological and behavioral alterations [2]. Although some authors argue that forelimbs can be used as an internal control [2], we would advise considering possible systemic effects of hindlimb unloading and being careful when applying it. It is also relevant to other immobilization techniques where one of the limbs remains "unaffected" and could be used as an internal control.

12.4 Casting, Bandaging, and Denervation

Different surgical techniques such as nerve or tendon ectomies have been used in the past for the reproduction of microgravity effects by induction of localized extremities disuse. For instance, commonly used sciatic neurectomy is a visually confirmed resection of 3–4 mm of sciatic nerve that leads to efficient denervation of all regions of the hindlimb [12]. These techniques lead to not only irreversible immobilization and significant bone loss and muscle atrophy but also multiple side effects.

Non-invasive methods such as casting, bandaging, and injection of Clostridium botulinum toxin have become more popular in the recent years. Clostridium botulinum toxin type A is a bacterial metalloprotease causing muscle paralysis and therefore limb disuse by the inhibition of neurotransmitter release. The injection is done into the posterior lower limb musculature. The major advantage of this technique compared to neurectomy is non-invasiveness and possibility of complete recovery within several months [13].

Bandaging and casting are methods of immobilization based on fixation of extremities in constant position by applying either elastic tape (bandaging) or hard orthopedic plaster (casting). During bandaging procedure, a hindlimb of anesthetized animal is immobilized against the abdomen with few layers of elastic bandages. Ankle joints and the knee are placed in extension, and the hip joint is placed in flexion [14]. The immobilized limb should not touch the floor of the cage during animal's movement. The gravitational loading, normally distributed between both hindlimbs, rests on the free limb. Animals are free to move and can easily reach food and water. The bandage should be examined daily and replaced twice a week [14].

Casting is very similar to bandaging but it allows to fix animal's extremities in desirable positions with precise adjustment of joint angles and muscle straitening. This feature of casting techniques allows to get either plantarflexion or dorsiflexion cast immobilization. Dorsiflexion of the ankle joint at an angle of 35° by casting of a limb was used by Nemirovskaya [15] in her study of adaptation mechanisms to microgravity in combination with tail suspension model. Casting can be not only unilateral but also bilateral. This model was recommended as a reliable cast immobilization particularly for mice because small size of animals is a technical challenge [16]. Casting is performed on anesthetized animals. The cast covers both hindlimbs and the caudal fourth of the body. A thin layer of padding is recommended to be placed underneath the cast to prevent abrasions. To minimize freedom of movement of limbs, slight pressure should be applied when wrapping the casting tape. To resist the cast against chewing on, fiberglass material can be applied over the cast. The animals can move using their forelimbs to reach food and water. The mice should be monitored daily for abrasions, chewed plaster, venous occlusion, and problems with ambulation [16].

12.5 Conclusions

Nowadays, hindlimb unloading is the only technique which imitates more physiological alterations relevant to spaceflight than any other on-ground model. Social isolation is not a common case for experiments conducted in space but it is a significant source of stress for animals subjected to unloading. This aspect should be taken into account when comparing results from spaceflights to results from on-ground models. Future development of hindlimb unloading could help tackle this issue.

References

[1] Morey-Holton, Emily R. and Ruth K. Globus. "Hindlimb Unloading of Growing Rats: A Model for Predicting Skeletal Changes During Space Flight." *Bone* 22, no. 5 (1998): 83–88.

[2] Morey-Holton, Emily R. and Ruth K. Globus. "Hindlimb Unloading Rodent Model: Technical Aspects." *Journal of Applied Physiology* 92, no. 4 (2002): 1367–1377.

[3] Dupont, Erwan, Laurence Stevens, Laetitia Cochon, Maurice Falempin, Bruno Bastide, and Marie-Helene Canu. "ERK is Involved in the Reorganization of Somatosensory Cortical Maps in Adult Rats Submitted to Hindlimb Unloading." *PLoS ONE* 6, no. 3 (2011): e17564.

[4] Aviles, Hernan, Tesfaye Belay, Monique Vance and Gerald Sonnenfeld. "Effects of Space Flight Conditions on the Function of the Immune System and Catecholamine Production Simulated in a Rodent Model of Hindlimb Unloading." *Neuroimmunomodulation* 12 (2005): 173–181.

[5] Tsvirkun, Darya, Jennifer Bourreau, Aurélie Mieuset, Florian Garo, Olga Vinogradova, Irina Larina, Nastassia Navasiolava, Guillemette Gauquelin-Koch, Claude Gharib and Marc-Antoine Custaud. "Contribution of Social Isolation, Restraint, and Hindlimb Unloading to Changes in Hemodynamic Parameters and Motion Activity in Rats." *PLoS ONE* 7, no. 7 (2012): e39923.

[6] Riley, Danny A., Glenn R. Slocum, James L. Bain, Frank R. Sedlak and James W. Mellender. "Rat Hindlimb Unloading: Soleus Histochemistry, Ultrastructure, and Electromyography." *Journal of Applied Physiology* 69 (1990): 58–66.

[7] Ferreira, Andries J., Jacqueline M. Crissey and Marybeth Brown. "An Alternant Method to the Traditional Nasa Hindlimb Unloading Model in Mice." *Journal of Visualized Experiments* no. 49 (2011): e2467.

[8] Knox, Micheal, James D. Fluckey, Patrick Bennett, Charlotte A. Peterson and Esther E. Dupont-Versteegden. "Hindlimb Unloading in Adult Rats Using an Alternative Tail Harness Design." *Aviation, Space, and Environmental Medicine* 75, no. 8 (2004): 692–696.

[9] Wagner, Erika B., Nicholas P. Granzella, Hiroaki Saito, Dava J. Newman, Laurence R. Young and Mary L. Bouxsein. "Partial Weight Suspension: A Novel Murine Model for Investigating Adaptation to Reduced Musculoskeletal Loading." *Journal of Applied Physiology* 109, no. 2 (2010): 350–357.

[10] Halloran, Bernard P., Daniel D. Bikle, Charlotte M. Cone, Emily Morey-Holton. "Glucocorticoids and Inhibition of Bone Formation Induced by Skeletal Unloading." *American Journal of Physiology* Endocrinology and Metabolism 255 (1988): 875–879.

[11] Blottner, Dieter, Najet Serradj, Michele Salanova, Chadi Touma, Rupert Palme, Mitchell Silva, Jean Marie Aerts, Daniel Berckmans, Laurence Vico, Yi Liu, Alessandra Giuliani, Franco Rustichelli, Ranieri Cancedda and Marc Jamon. "Morphological, Physiological and Behavioural Evaluation of a 'Mice in Space' Housing System." *Journal of Comparative Physiology* 179, no. 4 (2009): 519–533.

[12] Kodama, Yoshiaki, Peter Dimai, John E. Wergedal, Matilda Sheng, Rashmi Malpe, Stepan Kutilek, Wesley Beamer, Leah R. Donahue, Clifford J. Rosen, David J. Baylink and John R. Farley. "Cortical Tibial Bone Volume in Two Strains of Mice: Effects of Sciatic Neurectomy and Genetic Regulation of Bone Response to Mechanical Loading." *Bone* 25, no. 2 (1999): 183–190.

[13] Manske, Sarah L., Steven K. Boyd and Ronald F. Zernicke. "Muscle and Bone Follow Similar Temporal Patterns of Recovery from Muscle-Induced Disuse Due to Botulinum Toxin Injection. *Bone* 1, no. 46 (2010): 24–31.

[14] Ma, Yanfei, Webster S.S. Jee, Zhongzhi Yuan., Wei Wei, Hongka Chen, Sunwah Pun, Haohai Liang and Chaohua Lin. "Parathyroid Hormone and Mechanical Usage have a Synergistic Effect in Rat Tibial Diaphyseal Cortical Bone." *Journal of Bone and Mineral Research* 14 (1999): 439–448.

[15] Nemirovskaya, Tatiana L., Boris S. Shenkman, Anna Muchina, Yaroslav Volodkovich, Maria Sayapina, Olesya Larina and Elena Bratcseva. "Role of Afferent Control in Maintaining Structural and Metabolic Characteristics of Stretched Soleus in Rats Exposed to Hindlimb Suspension." *Journal of Gravitational Physiology* 9, no. 1 (2002): 121–122.

[16] Frimel, Tiffany N., Kapadia Fatema, Gabriel S. Gaidosh, Ye Li, Glenn A. Walter and Krista Vandenborne. "A Model of Muscle Atrophy Using Cast Immobilization in Mice." *Muscle Nerve* 32 (2005): 672–674.

13

Human: Bed Rest/Head-Down-Tilt/ Hypokinesia

Marie-Pierre Bareille and Alain Maillet

MEDES, Toulouse, France

13.1 Introduction

Many space agencies and in some cases even individual investigator teams around the world are involved in organizing bed-rest studies. However, the conditions in which these studies are performed are quite diverse. Differences lie, for example, not only in the organization and the environmental conditions of studies (duration of studies, angle of the bed, sunlight exposure, sleep/wake cycles, nutritional standards and control, etc.) but also in the scientific measurements taken. Indeed, like in a spaceflight, several scientific experiments are always carried out in the framework of ground-based studies. Thus, the scientific results of such studies may be affected by all these factors, and this complicates drawing overall conclusions and comparing results between the different studies. Furthermore, the experimental conditions are not always fully detailed in the scientific publications, and sometimes the authors report their results as if their experiment was the only one conducted in the study.

Therefore, in the past ten years, efforts were gradually made by the space agencies, the teams conducting bed-rest studies and the scientists to standardize as far as possible the design, the format, support and conduct of the studies. Recently, in order to achieve better standardization of bed-rest studies in the spaceflight context, an International Academy of Astronauts (IAA) study group was initiated, including members from most of the entities who are actively pursuing this type of activity.

This chapter focuses on the study design and logistics of bed-rest studies as used in the main facilities conducting studies around the world trying to emphasize the factors affecting the results and why.

13.2 Experimental Models to Mimic Weightlessness

The most used methods to simulate microgravity on Earth include immersion, bed rest, chair rest, isolation, hyperbaric environments and immobilization of animals. None of these techniques precisely duplicate near weightlessness because gravity cannot be entirely eliminated on Earth. However, two separate approaches, head-out water immersion and bed rest, have provided possibilities for long-term exposures and produce changes in body composition (including body fluid redistribution) and cardiovascular and skeletal muscle characteristics that resemble the effects of microgravity [1]. The common physiological denominator is the combination of a cephalad shift of body fluids and reduced physical activity.

13.2.1 Bed Rest or Head-Down Bed Rest?

Toward the end of the 1960s, Soviet investigators evolved a new method of bed rest in which the subject was positioned with the head lower than the feet, rather than horizontal, after having analyzed subjective comments of cosmonauts received after flight. Indeed, they complained to their medical staff that on their return from space they had a hard time sleeping because they had the sensation that they were slipping off the foot of the bed. They tried to correct the situation by raising the foot of the bed until it felt horizontal and they could get back to sleep. Every night they lowered the foot of the bed a little until lying horizontal felt normal again. They also suggested that the head-down-tilt position more closely reproduced the feelings of head fullness and awareness experienced during flights. Russian researchers took note of this observation and surmised that perhaps the head-down position on Earth was closer to what it felt like to be in space. The head-down bed-rest (HDBR) simulation model was born (Figure 13.1). The first study compared responses from horizontal bed rest and –4° head-down tilt. Since then, additional studies have been conducted with head-down positions ranging from –2° to –15° and lasting for 24 hours to 370 days [2]. In general, head-down bed rest induces findings more rapidly and profoundly than its horizontal counterpart [1]. The Soviets tested –15°, –10° and –5° for comfort, acceptability and magnitude of response and decided –6° was the best compromise. In many ways, HDBR made it in fact more comfortable for the subjects. They could lean over the side of the bed to eat. They could raise their knees as well since that only increased the head-ward fluid shift. In 1977/1978 joint USA/USSR 7-day studies done both at the IMBP (Institute

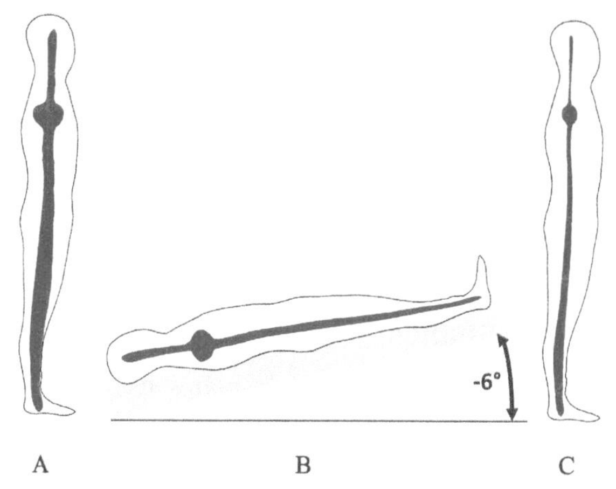

Figure 13.1 This shows the fluid shift from the lower to the upper part of the body induced by bed rest. (A) On Earth (1 g), the main part of the blood is located in the legs. (B) In Head-down bed rest (–6°), the thoraco-cephalic fluid shift stimulates central volume carotid, aortic and cardiac receptors inducing an increase in diuresis and natriuresis and a decrease in plasma volume. (C) While standing, this venous part of the blood falls to the lower part of the body (abdomen and legs). To come back to the heart, the blood has to go against the gravity. In that case, less blood comes back to the heart, the blood pressure tends to decrease. As in spaceflight, cardiovascular deconditioning characterized by orthostatic intolerance is observed at the end of bed rest.

for BioMedical Problems) in Moscow and at ARC (Ames Research Centre) in Moffett Field, CA, compared HDBR with horizontal bed rest and confirmed the added value of HDBR [3].

13.2.2 Immersion and Dry Immersion

With the advent of human spaceflight in 1961, immersion in water was used as a logical model for reducing the pull of gravity on the mass of the body. But it soon proved impractical because remaining in water for more than a day brought on unpleasant consequences. Therefore, this experimental model was used mainly for short-term studies and especially for fluid regulation [4–6].

Because the Soviets wanted to investigate on the long term in preparation of their spaceflights, they developed the dry immersion model. This model gave the possibility to the IBMP to conduct studies up to 57 days [2, 7]. Only a few number of these dry immersion systems were existing in Russia; today IBMP and a private company propose these systems for some rehabilitation centers [8]. A recent collaboration between the French Space Agency (CNES) and IBMP was initiated, and two of these systems have been installed at MEDES (Toulouse, France) to develop this experimental model in Europe and allow comparison with head-down bed rest. A first experiment is planned for 2015.

13.3 Overall Design of the Studies

13.3.1 Duration of the Studies

Since the end of the 1960s, many bed rest studies have been conducted lasting from several hours to a maximum of 370 days. The shortest studies are appropriate to investigate cardiovascular changes (4 hours to 1 week) while longer-term studies are needed to study other physiological changes, in particular bone loss and muscle atrophy (minimum of three weeks).

In the United States, the long-term studies were of 60, 70 or 90 d. In Russia, 120-day bed-rest studies have also been conducted and of course the 370 days of the Moscow study went way beyond anything that had been done before and since then.

After the first long-term bed-rest experiment performed in Europe in 2001–2002 (i.e. 90 days of head-down bed rest with human subjects), the duration of the bed-rest studies has been standardized to 5 d for short-term, 21 d for medium-term and 60 d for long-term studies.

The IAA Study Group has started writing guidelines for the standardization of bed-rest studies. The length of bed rest will be categorized into 3 different durations with associated pre- and post-bed-rest phases. The table below describes these categories.

The bed-rest studies are divided into three phases: (1) a pre-bed-rest phase for acclimation and baseline data collection, (2) a bed rest phase, and (3) a

Table 13.1 Categories for bed-rest study duration

Category	Pre-Bed-Rest Baseline Data Collection (BDC)	Head-Down-Tilt (HDT)	Post-Bed-Rest Recovery (R)
Short-Duration Bed Rest	5–7 days	5–14 days	3–6 days
Medium-Duration Bed Rest	7–14 days	15–59 days	7–14 days
Long-Duration Bed Rest	14 days (or more)	60 days (or more)	14 days (or more)

post-bed-rest recovery phase for post-bed-rest testing and reconditioning. For scheduling consistency, each study day is referred to with a conventional naming system. Pre-bed-rest days began at BR-X and ended on BR-1. Days in bed rest began at BR1. Post-bed-rest days began at BR +0 and subjects were released on BR +X.

Obviously, the standardization of the bed-rest experiment is of major importance both for the bed-rest phase and for the control phase (i.e. pre- and post-bed-rest phases).

The pre-bed-rest phase to collect baseline data is of particular importance because the deconditioning of the subjects can start during the pre-bed-rest phase. Indeed, keeping the subjects in normal daily life condition is really challenging, and significant changes can occur among subjects if habits are changed too much (mainly in term of exercise level and diet). The goal of this pre-bed-rest period is to homogenize the subjects' pool.

13.3.2 Design of the Bed-Rest Studies

The long-term bed-rest studies use obviously a parallel design, and the medium- or short-term bed-rest studies can have a parallel or a crossover design. Of course, in that case the influence of confounding covariates is reduced because each crossover subject serves as his/her own control, and crossover designs require fewer subjects than do non-crossover designs. Nevertheless, the washout period between the two treatment periods has to be carefully evaluated, and the planning for sufficiently long washout periods does require expert knowledge of the dynamics of the recovery. Then, for a test subject, a study with a crossover design is longer and the risk of withdrawal is higher.

13.3.3 Number of Volunteers

Determining the optimal sample size for a study assures an adequate power to detect statistical significance. Hence, it is a critical step in the design of a planned research protocol. Using too many participants in a study is expensive and exposes more number of subjects to the procedure. Similarly, if the study is underpowered, it will be statistically inconclusive and may make the whole protocol a failure. In addition, even in well-designed and well-conducted studies, it is unusual to finish with a dataset, which is complete for all the subjects recruited, in a usable format. The reason could be subject factors like subjects may fail to particular questions, physical measurements may suffer from technical problems, and in high-demanding studies dropouts before the study ends are not unlikely.

Calculation of sample size requires precise specification of the primary hypothesis of the study and the method of analysis. Usually, several protocols are implemented on the same bed-rest study and this calculation cannot often be done. So the bed-rest studies include usually 8–12 subjects per group. There is always a control group used as a standard for comparison and one or two intervention groups depending on the number of countermeasures to be tested. The sample size in each group at the end of the study is of course positively correlated with the statistical power of the study. In total, the number of subjects for a study with one control and one intervention group is from 16 to 24 (i.e. 8–12 per group). Thus, a larger sample size at the beginning of the study gives a better chance to keep greater power, especially in high-demanding studies like long-term bed-rest studies or in crossover studies running on a long time with a higher risk of dropouts.

13.3.4 Number of Protocols

What is specific in bed-rest studies in comparison with classical clinical trials is the number of protocols implemented on the same study. Indeed, bed-rest studies are complex, high-demanding and expensive studies. Usually several protocols investigating different fields are selected to be implemented on the same study. The planning is then a crucial issue because it is usually very tight and the interferences between the different tests have to be anticipated and avoided as far as possible. But amazingly, publications report the results of each experiment as if the experiment was the only one implemented on the study. The hypothesis that some results may have been affected by some other tests or other activities of the subjects is almost never raised.

13.3.5 Selection Criteria

Of course the general selection criteria to be included in a bed-rest study are in principle the same for all the studies wherever they are conducted. Potential subjects have to be healthy and should not have any history of cardiovascular or other major diseases, and all undergo an extensive medical examination before being included in the study. Nevertheless, there are sometimes some discrepancies in the main criteria like the range of ages or body mass index (BMI) which have a matter of particular importance. The studies carried out in the United States screen people aged 24–55 to match the age of most astronauts, while in Europe the range was 25–45 and now 20–45. Regarding the BMI, the studies carried out in the United States screen people with a BMI between 21 and 30 and in Europe the range was 20–26.

In the international guidelines for standardization of bed-rest studies, the general age range is now 20–55 years and it is stated that smaller ranges should be defined prior to each study. The BMI has been fixed to 20–26 kg/m^2. The international recommendation for the fitness level remains vague: "in general, it should be defined which activity level/fitness level the subjects should have or should not have. There is a high variability of fitness levels between different individuals". Indeed this decision may be more or less important depending on the main research question for each study. However, it should be taken into account that the performance level itself prior to the study as well as the ability to adapt to training/detraining may have a significant impact on the results of a study, especially in the context of bed rest. It is therefore mandatory, besides the decision whether trained or untrained subjects will be needed for the study, to select a group of test subjects that is as homogeneous as possible regarding their fitness level.

13.4 Directives for Bed Rest (Start and End of Bed Rest, Conditions During Bed Rest)

13.4.1 Respect and Control of HDT Position

The conditions of the bed-rest phase and how the –6° HDT position is maintained and controlled could also give rise to variability. Some reports state that the subjects were allowed to use the bathroom or a bedside pot—standing from squatting is an excellent orthostatic test—or sat on a bedpan on the bed, presumably as an acceptable compromise.

13.4.2 Activity Monitoring of Test Subjects

To document compliance of the subjects with the requirements, their activity shall be monitored. This should be done by redundant methods such as video control, pressure sensors, by subject monitors in person or activity measurements by an actimeter. Activity during the bed-rest period can also be recorded with telemetric electromyography on randomized study days.

For video control, the rooms of the subjects are equipped with video cameras. Video recording shall be continuous. To respect the subject privacy, the cameras should point only to their upper body. The period of non-video control should be documented in a specific form.

13.4.3 First Day of Bed Rest

It had been the custom among those running bed-rest studies in the early days to call 'bed-rest day one' (BR1) a day when subjects did not get out of bed after waking up but continued to stay in bed. This produced a very gradual response. When HDBR studies became popular, subjects were allowed to get out of bed in the morning, shower and eat breakfast while the foot of the bed was raised producing the head-down angle of –6° vs. horizontal position. Subjects would then go back to bed at 9 am to begin BR1. This routine produces a maximum possible posture change (1–0 Gz), inducing a full head-ward fluid shift that triggers a significant and consistent sequence of events lasting about 24 hours, leading subsequently to all the changes we have become so familiar with. If one stays in bed on awakening on BR1, and the foot of the bed is merely raised to the –6° angle, the initial physiological response is dampened. No significant cardiovascular and endocrine changes occur during the first 24 hours of bed rest that would normally accompany the maximal postural change [3].

How the first day of BR (BR1) begins is frequently not mentioned in published paper methodology, but can make all the difference to the time course and magnitude of changes. This is particularly true if the duration of the study is relatively short. There are very good scientific reasons (i.e. for fluid shift and initial volume regulation) for using HDBR especially if on BR1 the subject goes from standing upright to head down, in other words uses the maximum postural change to mark the beginning of bed rest [9–11].

13.4.4 Physiotherapy

Immobilization through bed rest can cause side effects, for example, neck pain, back pain or headaches. Physiotherapy can provide relief in some cases and should be considered by the medical doctors as treatment before using drugs. The aim of these massages is to prevent the muscular pain and the occurrence of thrombophlebitis. These massages will last 30 minutes and will be planned in order to avoid interference with the scientific protocols (intensity and frequency of these massages should be carefully handled in order not to be considered as countermeasures).

If allowed, stretching regimen or deep breathing exercises which are one of the most effective and easiest techniques of self-regulation have to be carefully monitored because they can respectively affect muscle and cardiovascular testing.

13.5 Operational/Environmental Conditions

13.5.1 Housing Conditions and Social Environment

Housing conditions are site- or protocol specific depending on the facility infrastructure: Participants can be accommodated one or more by room; in this latter case, subjects are matched as roommate pairs (or more) upon psychological criteria.

The social environment, visitors, communication with the outside world and access to news can all become sources of comfort or irritation. During BR studies, the participants are usually allowed to freely communicate with each other, to watch television and video, to listen to radio, to read books and magazines, to work on computer and to use the Internet. The main difference between the studies regarding the social environment is based on the possibility to receive visitors or not. In the studies conducted in Europe, at MEDES and at DLR as well, visitors (family or friends) are not allowed for the obvious reasons of the impact on the mood and psychology of the volunteers and the disparity it could create between them. Confinement and the environment could alter the physiological baseline through anxiety, loneliness, unwanted interaction with staff or strangers or lack of privacy. All or any of these could affect the results [3].

The other environmental conditions like temperature, pressure and humidity are controlled and maintained to allow physiological comfort of the test subjects.

13.5.2 Sunlight Exposure, Sleep/Wake Cycles

Subjects are expected to wake at lights on and to cease activity at lights out. Depending on the facility, the standard sleep/wake schedule lights are turned on at 06:00 or 07:00 and are turned off at 22:00 or 23:00. Nevertheless, sometimes it is necessary to interrupt the sleep period for early-morning or late-evening test procedures. Light/dark cycles can affect circadian rhythms and therefore result in lack of sleep as well.

The exposure to daylight should be controlled as it elicits physiological reactions in the human body that may influence the results of bed-rest studies. Especially if study campaigns take place during different seasons, different exposure to daylight may jeopardize the results of the study. The exclusive use of artificial light would be the easiest way to control daylight. However then, supplementation of Vitamin D would be mandatory.

13.5.3 Diet

Dietary consistency and control is of paramount importance to the reliability of the results. Performance conditions of bed-rest protocols depend on the investigations performed (diet was sometimes controlled or sometimes just monitored, depending on whether nutrition protocols were included or not). Some studies strictly control diet, others feed subjects ad libitum.

Because nutrition is the source of energy and substrates used as precursors for synthesizing the functional body units (cells and their core constituents, macromolecules, proteins, DNA, etc.), and the numerous co-factors, such as micronutrients, to support enzyme activity and detoxification/repair mechanisms, nutrition is central to the functioning of the body. Conversely, poor nutrition can compromise many of the physiological systems and also mood and performance [12]. Dietary consistency and control is of paramount importance to the reliability of the results of any clinical studies. This is obvious not only for metabolic studies but also for bone, muscle and cardiovascular studies. First, the choice of the ratio of macronutrients is of crucial importance during the control pre-bed-rest period to maintain the volunteers in situations close to their usual daily life conditions. To achieve this aim, it is important that the volunteers do not have very different dietary habits. The macronutrient composition of the diet also influences the bed-rest outcomes. In a crossover design, 60-h bed rest in eight males (2 days of washout) with either a high-carbohydrate diet (70% of energy intake) or a high-saturated-fat diet (45% of energy intake as fat and 60% of saturated fatty acid) [13] showed that insulin sensitivity decreased by 24% with the high-saturated-fat diet but did not change with the high-carbohydrate diet.

Until recently in most HDBR experiments, energy intake was adjusted so that the body mass was clamped to the pre-bed-rest values [14, 15]. Of course during head-down bed rest (HDBR), body composition is altered [16], muscle mass decreases due to disuse and fat mass varies according to diet prescription. The mass clamping approach leads, however, to a positive energy balance and an increase in fat mass without changes in body mass [14, 17]. A positive energy balance due to overfeeding is a confounding factor that exaggerates the deleterious effects of physical inactivity, and the effects of overfeeding cannot be dissociated from those of simulated weightlessness. Indeed positive energy balance during inactivity is also associated with greater muscle atrophy and with activation of systemic inflammation and antioxidant defense [18, 19]. Other observations suggest that nutrition may also play a role in the cardiovascular deconditioning syndrome as observed in Muslim army pilots.

An adequate nutrient supply to accurately derive the true effects of bed rest alone is of particular importance. In order to standardize bed-rest experiments in a way that controls energy balance, one needs to adjust in real time energy intake to energy expenditure. Two long-term bed-rest studies [20, 21] during which the diet was tightly controlled confirmed that clamping fat mass may be possible [22], yet more technical precision is needed. Nevertheless, measuring daily energy expenditure remains a technical challenge, particularly when an exercise protocol is chosen to counteract any muscle and bone atrophy because it is more difficult to match total energy expenditure. The objective of a very recent study conducted by European researchers at MEDES (Toulouse, France) is to validate the minimum set of techniques mandatory to match total energy expenditure in future studies including physical exercise.

13.5.4 Testing Conditions

Considering how important the elimination of posture change by bed rest is to the fidelity of the results to microgravity, it is amazing that published papers often do not mention the position in which tests are performed. Closer scrutiny may reveal that plasma volume was measured in the seated position, or that subjects were allowed to use the bathroom [3].

13.5.5 Medications

Other factors that may interfere with the results of a bed-rest study but that are transparent to an investigator include clinical aspects of a study. For instance, a mild laxative may routinely be prescribed. Those that work by drawing fluid into the gut to soften stools may well interfere with the results. Headaches often of a sinus nature are not uncommon. Headaches may also be triggered by overhead lighting.

References

[1] Sandler, H., and Vernikos, J. *Inactivity: Physiological effects*. Orlando: Academic Press, 1986.

[2] Atkov, O.Y., and V.S. Bednenko. *Hypokinesia and Weightlessness: Clinical and Physiologic Aspects*. Madison: International Universities Press, Inc,, 1992.

[3] Pavy-Le Traon, A., M. Heer, M.V. Narici, J. Rittweger, and J. Vernikos. "From Space to Earth: Advances in Human Physiology from 20 years of Bed Rest Studies (1986–2006)." *European Journal of Applied Physiology* 101, no. 2 (2007): 143–194.
[4] Epstein, M. "Renal Effects of Head-Out Water Immersion in Humans: Implications for an Understanding of Volume Homeostasis." *Physiological Reviews* 58 (1978): 529–581.
[5] Epstein, M. "Renal Effects of Head-Out Water Immersion in Humans: a 15-year Update." *Physiological Reviews* 72, no. 2 (1992): 563–621.
[6] Norsk, P. "Gravitational Stress and Volume Regulation." *Clinical Physiology* 12 (1992): 505–526.
[7] Koryak, Y. "Mechanical and Electrical Changes in Human Muscle After Dry Immersion." *European Journal of Applied Physiology* 74 (1996): 133–140.
[8] Navasiolava, N.M., et al. "Long-term dry immersion: review and prospects." *European Journal of Applied Physiology* 111, no. 7 (2011): 1235–1260.
[9] Maillet, A., et al. "Hormone Changes Induced by 37.5-h Head-Down Tilt (–6°) in humans." *European Journal of Applied Physiology*, 68 (1994): 497–503.
[10] Hughson, R.L., et al. "Investigation of Hormonal Effects During 10-h Head Down Tilt on Heart Rate and Blood Pressure Variability." *Journal of Applied Physiology* 78, no. 2 (1995): 583–596.
[11] Diridollou, S., et al. "Characterisation of Gravity-Induced Facial Skin Oedema Using Biophysical Measurement Techniques." *Skin Research and Technology* 6 (2000): 118–127.
[12] Blanc, S., et al. "THESEUS–Cluster 1: Integrated Systems Physiology-Report. Strasbourg, EC FP7 Grant 242482, 2012.
[13] Stettler, R., et al. Interaction between dietary lipids and physical inactivity on insulin sensitivity and on intramyocellular lipids in healthy men. *Diabetes Care* 28, 6 (2005):1404–1409.
[14] Gretebeck, R.J., D.A. Schoeller, E.K. Gibson, and H.W. Lane. "Energy Expenditure During Antiorthostatic Bed Rest (simulated microgravity)." *Journal of Applied Physiology* 78, no. 6 (1995): 2207–2211.
[15] Bergouignan, A., F. Rudwill, C. Simon, and S. Blanc. "Physical Inactivity as the Culprit of Metabolic Inflexibility: Evidence from Bed-Rest Studies." *Journal of Applied Physiology* 111 (2011): 1201–12010.

[16] Stein, P.T. "The Relationship Between Dietary Intake, Exercise, Energy Balance and the Spacecraft Environment." *Pflugers Archiv* 441, no. 2–3 Suppl. (2000): R21–R31.

[17] Krebs, J.M., V.S. Schneider, H. Evans, M.-C. Kuo, and A.D. LeBlanc. "Energy absorption, lean body mass, and total body fat changes during 5 weeks of continuous bed rest." *Aviation Space and Environmental Medicine* 61, no. 4 (1990): 314–318.

[18] Biolo, G., et al. "Calorie Restriction Accelerates the Catabolism of Lean Body Mass During 2 Week of Bed Rest." *American Journal of Clinical Nutrition* 82, no. 2 (2007): 366–372.

[19] Biolo, G., et al. "Positive Energy Balance is Associated with Accelerated Muscle Atrophy and Increased Erythrocyte Glutathione Turnover During 5 Week of Bed Rest." *American Journal of Clinical Nutrition* 88 (2008): 950–958.

[20] Bergouignan, A., et al. "Effect of Physical Inactivity on the Oxidation of Saturated and Monounsaturated Dietary Fatty Acids: Results of a Randomized Trial." *PLoSClin Trials* 1 (2006): e27.

[21] Bergouignan, A., et al. "Physical Inactivity Differentially Alters Dietary Oleate and Palmitate Trafficking." *Diabetes* 58 (2009): 367–376.

[22] Bergouignan, A., et al. "Regulation of Energy Balance During Long-Term Physical Inactivity Induced by Bed Rest with and Without Exercise Training." *Journal of Clinical Endocrinology and Metabolism* 95 (2010): 1045–1053.

14

Clinostats and Other Rotating Systems—Design, Function, and Limitations

Karl H. Hasenstein[1] and Jack J. W. A. van Loon[2]

[1]University of Louisiana at Lafayette, Lafayette, LA 70504, USA
[2]VUmc, VU-University, Amsterdam, The Netherlands

14.1 Introduction

Clinostats are rotational devices that have been in use ever since Julius Sachs invented a clockwork-driven device that rotated growing plants around their growth axis at the end of the 19th century [1]. The initial clinostat systems were mostly used in plant studies and rotated with a relatively slow frequency on the order of one rotation per couple of hours up to about 10 revolutions per minute. Seeds or adult plants were fixed to the clinostat within some semisolid or soil substrates. Although mostly used to simulate microgravity, there are some interesting adaptations of these systems made over the years. For instance, clinorotation was combined with centrifugation to generate a partial gravity in order to establish the gravity threshold of various systems [2–4]. In the 1960s, Briegleb [5] introduced the concept of a fast-rotating (on the order of 60 rpm) clinostat dedicated for liquid cell culture studies. Other improvements were implemented over time, and the most recent modification resulted in the so-called 3D clinostat [6] and the random positioning machine (RPM) [7, 8]. The widespread use of clinostats and the often-found notation that all it takes to eliminate gravitational effects on organism is to rotate them prompt this description of the purpose, goals, and limits of these devices. The most important message may well be the simple statement that clinostats, albeit intended for this purpose, do not simulate microgravity. The detailed assumptions and consequences of rotational movements in a static force field, such as Earth's gravity, are often overlooked and may result in questionable or downright incorrect statements. Since clinostats can also be used to mimic fractional gravitational loads, their usefulness goes beyond averaging the gravity vector over time; they can also be used to study the effects of hypogravity.

14.2 Traditional Use of Clinostats

The most important constraints have been recognized early on and were identified as centrifugal forces ($Z = r\omega^2$, product of radius and angular velocity), phase shifts of mobile particles as a consequence of rotation ($\tan \phi = -\omega$ m/f, the offset between the angular displacement of the rotating structure and the rotation of a cellular particle), and friction experienced by a mass as a result of rotation ($V_f = g/((f/m)^2 + \omega^2)^{1/2}$, the sedimentation of a particle as a function of its mass and the viscosity of the medium) [9]. In addition, the direction of rotation is relevant, and related concerns led to the development of random positioning machines whose sole purpose is to not generate constant forces in any particular direction. Let us consider the significance of each of these parameters.

14.3 Direction of Rotation

The intended averaging of the gravity vector can be accomplished by rotating an object such as a plant, bacterium, or (small) animal around a horizontal axis. However, the direction of the object is also important. While traditionally plants were rotated around their longitudinal axis (i.e., the shoot–root axis is positioned horizontally), gravity averaging is also possible by rotating the long (i.e., shoot–root) axis of a plant perpendicularly to a horizontal axis (aka vertical clinorotation). While the extended size of this axis limits studies to relatively short seedlings, studies have shown that plants are more sensitive to this latter type of rotation as the growth rate decreases more than after horizontal clinorotation [10]. Thus, the direction of rotation affects the physiology of biological objects, especially when the biological objects themselves are rotating such as tendrils or circumnutating stems [11].

14.4 Rate of Rotation

Rotating a physical object such as a growing plant regardless of the orientation relative to the horizontal axis of rotation shows that the radius is not constant. Horizontal rotation affects stems to a lesser extent than leaves that have a larger radius than stems. Thus, the centrifugal force varies within the organism based on the distance from the rotational axis. If a biological object (plant, seedling,

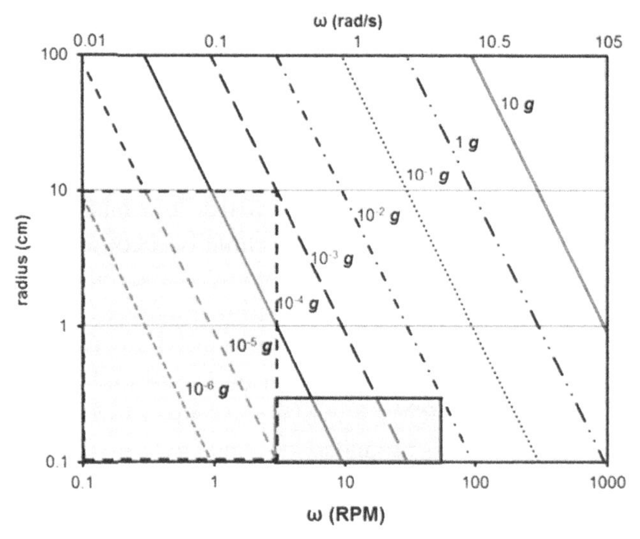

Figure 14.1 Log/log plot of radius and angular velocity (expressed as revolutions per minute and radians per second). The different lines define the centrifugal force induced by the respective rate of rotation. The rectangles exemplify usable dimensions and angular velocity ranges for slow-rotating (gray) and fast-rotating (blue) clinostats. The relative acceleration based on angular velocity and radius is shown as g-equivalent.

or other organism) is offset from the rotational axis, the centrifugal force is minimized (zero) only for structures that are aligned with the rotational axis. However, because gravity may be averaged but is never eliminated, elastic bending of the stem or leaves ensues and the center of rotation constantly shifts. Thus, the changing weight distribution causes bending stress and non-random mechanostimulation because structures such as leaves and petioles have a specific load-bearing design. A similar process might occur with mammalian cells [8]. However, simplifying the complexities of body structure and adjusting the rotational speed such that even larger dimensions are not exposed to more than the minimal centrifugal force require a careful consideration of applicable rotational speeds (Figure 14.1).

14.5 Fast- and Slow-Rotating Clinostats

While the slow-rotating clinostat simply considers the overall geometry and establishes a rotational regiment that fulfills predetermined conditions (e.g., centrifugal forces less than 10^{-3}g, Figure 14.1), the fast-rotating clinostat also considers the path of sedimentation in a fluid, typically an aqueous growth medium for small (<1 mm) organisms. In aqueous conditions, sedimentation and (slow) rotation result in appreciable side effects such as spiral combinations of movements stemming from sedimentation, centrifugation, and viscosity-dependent Coriolis force $g_f = 2\omega\varphi$, where g is gravity, ω is the angular velocity, and φ is the angle per unit time. When the frequency of rotation is increased, sedimentation of a particle will be less than the movement of the liquid, leading to a reduced radius that eventually becomes smaller than the size of the particle or a cell. Under those conditions, the rotation stabilizes the fluid around the particle, effectively eliminating gravity effects [5, 8, 12].

Related to the fast-rotating clinostat as introduced by Briegleb is the rotating wall vessel (RWV), which is mostly used in cell biology and tissue engineering [13]. Basically, the RWV is a relatively large (5–20 cm diameter) liquid-filled container that rotates around a horizontal axis at 10–20 rpm [14]. Samples within the container are prevented from settling by matching the rotation speed to the sedimentation velocity of the sample. This velocity depends on the specific density of the sample (cells, nodules, or others), their volume and shape, as well as the density and viscosity of the suspending medium. In the fast-rotating clinostat, cells rotate around their own center and experience no direct fluid shear force; in contrast, cells and tissues in the RWV are constantly falling within the fluid. The sedimentation velocity and direction combined with the rotation of the fluid generate spiral trajectories within the vessel [15]. The samples' motion relative to the fluid generates shear forces on a particle surface ranging from 180 to 320 mPa for 50-μm particles [14], up to 780 mPa for 300-μm spherical particles [16]. These are significant shear forces compared to physiologically relevant shear values, for example, endothelial cell responses that are on the order of 500–1000 mPa [17, 18]. Also, when the mass distribution within the particle is anisotropic, the particle may retain its orientation with respect to the gravity vector which is to be avoided if one wants to simulate microgravity.

14.6 The Clinostat Dimension

Tradition subdivides clinostats into 1D, 2D, and 3D (three-dimensional) systems. Of course since even the 1D clinostat operates as a three-dimensional system, the traditional naming is totally incorrect. Rather than referring to dimensions, clinostats may be better described by the number of rotating axes. Thus, a one-axis (or 1D) clinostat rotates around one axis. The orientation of this axis determines its properties. Under the correct conditions (dimension and rotational speed), it compensates or averages the vectorial character of gravity. In a vertical position, it moves an object but the gravity vector is consistently experienced toward the basal (Earth-facing) side of the object. Positioning the axis at an angle in between the horizontal and vertical position results in a net proportion of Earth's gravity such that the experienced or virtual gravity g_v is equal to the sin α (the angle from the horizontal) times g. A simulation of Moon's gravity (1/6 g) therefore requires an angle of about 10 degrees; Mars' gravity ($\sim$0.37 g) could be simulated by tilting the axis by 22 degrees. The advantage of using a single-axis clinostat lies in its simplicity of establishing virtual conditions.

Things become considerably more complicated when more than one axis is employed. However, a true rendition of all spatial orientation requires rotation around the three axes of space; adapting aviation terminology, pitch, yaw, and roll represent rotation around the horizontal (x) axis, the vertical (y) axis, and the z axis, respectively. This definition only applies when the rotation center is located inside the object of interest. Under these conditions, movement around the three axes is described by the rotation matrix $R^* = Rx(\alpha)\ Ry(\beta)\ Rz(\gamma)$, where α, β, and γ represent pitch, yaw, and roll angles.

$$Rx(\alpha) = \begin{pmatrix} 1 & 0 & 0 \\ 0 & \cos\alpha & -\sin\alpha \\ 0 & \sin\alpha & \cos\alpha \end{pmatrix}, \quad Ry(\beta) = \begin{pmatrix} \cos\beta & 0 & \sin\beta \\ 0 & 1 & 0 \\ -\sin\beta & 0 & \cos\beta \end{pmatrix},$$

$$Rz(\gamma) = \begin{pmatrix} \cos\gamma & -\sin\gamma & 0 \\ \sin\gamma & \cos\gamma & 0 \\ 0 & 0 & 1 \end{pmatrix}$$

This complete set of operation is in contrast to the two axes used in the so-called 3D clinostat or the random positioning machine (RPM). So what is the justification of calling a two-axis system a 3D clinostat? The simplest explanation relates to the ability to observe the entire three-dimensional space by moving, for example, a camera (or head) around only two axes (up/down and left/right). This concept was illustrated using a "ball RPM," which consists of a sphere resting on three support points, one of which consists of a wheel that drives the sphere by rotating in one or several directions (e.g., vertical or horizontal). The remaining support points are passive but omnidirectionally rotating spheres [7]. Alternatively, two drive wheels rotate around the normal axis of the sphere at different rates. Either system can rotate the ball in one or more planes, thus simulating a single-axis clinostat or an RPM. Regardless of the drive mechanisms, the distance between the surface of the sphere and the rotational axis varies and thus centrifugal accelerations can change.

Engaging a second drive or rotating the single drive wheel adds motion around a second axis and the superposition leads to a movement that depends on the relative angular velocity. If both drives operate synchronously, the plane of rotation tilts by the cross product of $Rx(\alpha)$ and $Ry(\beta)$. If the drives operate asynchronously, the movement of the sphere becomes complex. If the speed of the drivers differs but is constant, a phase angle results, and the motion of the sphere relative to the axis of rotation and the angular velocity change. Of course, as long as the sequence of rotational changes is known, nothing about the motion is random and the motion is reproducible. True randomness requires changes in the velocity of the drives that are not reproducible. However, it is doubtful whether such subtle modifications can be perceived by biological systems on the background of constant motion. Therefore, the "random positioning machine" might better be renamed "variable positioning machine."

Different ratios of angular velocity lead to well-known Lissajous figures projected onto the surface of the sphere with the position of the trace = A $\sin(\omega_0 t)$ + B $\cos(\omega_0 t)$. The phase shift between these two parameters determines the "tilt" of a point on the sphere surface but any rotation will be uniform; thus, different levels of gravity can be obtained based on the extent of the phase shift; zero corresponding to a one-axis clinostat and a phase shift of 10 degrees simulates Moon's gravity, as explained above. If the frequencies change, seemingly random patterns emerge (Figure 14.2).

Current developments also look into generating partial gravity in 3D rotating systems like an RPM. Dutch Space (Leiden, the Netherlands) presented a software-controlled partial g RPM (European Low Gravity Research

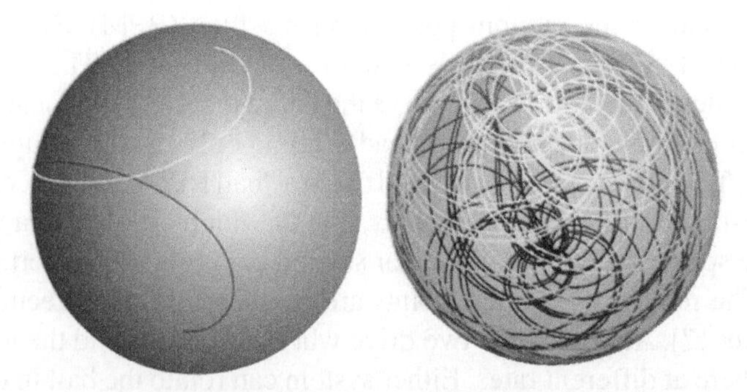

Figure 14.2 Projected traces of a surface point on a sphere that rotates with the same frequency for two perpendicular axes (left). Changing the frequency of one axis produces a distribution that covers the entire surface of the sphere. Calculations were performed after Kaurov [20].

Association in Vatican City in 2013) and a group from Switzerland published a comparable design [19].

14.7 Configurations of Axes

The previous example refers to two axes that rotate around a single point. However, the arrangement of rotational axes can be more complicated. For a single axis, no modifications are possible. Two rotational axes can be configured according to the description of the hypothetical sphere above; that is, they are part of a gimbal suspension. A second mode of arranging two axes consists of arranging the second axis perpendicular to the first axis of rotation. Such a device (Figure 14.3) has been implemented to examine the acceleration sensitivity of shoots and roots [4]. The relative rotation of the vertical axis of two wheels is used to turn "spokes" that extend from the center of the axis (Figure 14.3). This arrangement allows the entire system to function as a centrifuge if both upper and lower wheels rotate at the same angular velocity. If the lower wheel remains fixed and the top gear rotates, the experimental chambers that comprise the spokes rotate around their horizontal axis and are thus clinorotated as explained above for the 1-axis clinostat. Any additional movement by the lower wheel rotates the clinorotating chambers. This setup was used to determine the acceleration threshold of roots and shoots to about 10^{-3} and 10^{-4} g, respectively [4].

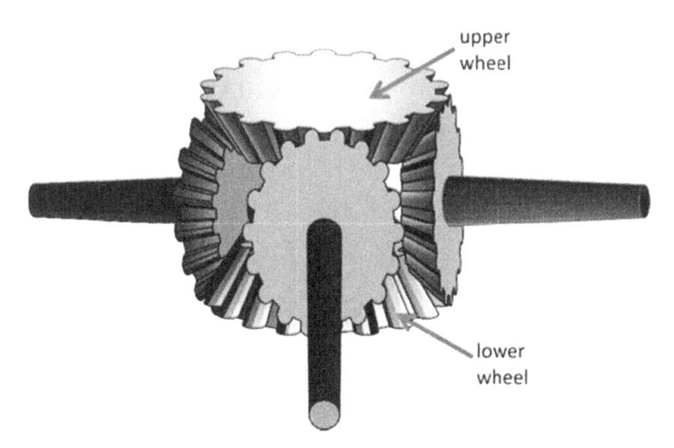

Figure 14.3 Drawing of a gearhead that translates the relative motion of a vertical shaft into a rotational motion of lateral axes. If the two center wheels rotate at the same rate, the horizontal axes function as a centrifuge and only yaw rotation applies. If the horizontal wheels spin at unequal rates, the lateral axes rotate and can drive a 1D clinostat with variable yaw and roll.

The creative arrangement of axes can be applied to the variable positioning system. If the lateral axes in Figure 14.3 contain a two-gimbal-supported suspension, a constant yaw acceleration is possible that averages all other motions relative to g. The constant yaw rotation generates a centrifugal force that is superimposed onto (1D) clinorotation. The effect of the angular velocity allows for the determination of a g-threshold value [4].

The data in Figure 14.1 indicate that the clinostatting of plants should not induce effects that are related to residual gravitational effects as long as the radius of rotation is less than the calculated values. However, recent studies clearly indicate that the rate of rotation over the magnitude of an octave (10-fold increase in frequency) affects induced curvature after gravistimulation. Brassica roots that were horizontally placed for 5 min and monitored for two hours of clinorotation between 0.5 and 5 rpm showed stronger curvature at higher frequency although the effective radius was less than 3 cm [10]. This observation, in addition to many others [21–25], indicates that the clinostat-associated mechanostimulation exerts largely unknown effects that prevent labeling clinorotation even in its most sophisticated form "microgravity simulation." The effects of mechanical unloading that are the hallmark of free-fall and orbital conditions do not apply to clinorotation. Nonetheless, the fascinating possibilities of manipulating organisms relative to the gravity vector for various times and under various conditions are bound to shed light on important aspects of sensory biology.

Acknowledgement

This work was supported by NASA grant NNX10AP91G to KHH and ESA grant TEC-MMG/2012/263 and Netherlands Space Office NSO/NWO grant to JvL.

References

[1] von Sachs, J. "Über Ausschliessung der geotropischen und heliotropischen Krümmungen wärend des Wachsthums." *Würzburger Arbeiten* 2 (1879): 209–225.

[2] Galland, P., H. Finger, and Y. Wallacher. "Gravitropism in Phycomyces: Threshold Determination on a Clinostat Centrifuge." *The Journal of Plant Physiology* 161 (2004): 733–739.

[3] Laurinavicius, R., D Svegzdiene, B Buchen, and A. Sievers. "Determination of the Threshold Acceleration for the Gravitropic

Stimulation of Cress Roots and Hypocotyls." *Advances in Space Research* 21 (1998): 1203–1207.

[4] Shen-Miller, J., R. Hinchman, and S.A. Gordon. "Thresholds for Georesponse to Acceleration in Gravity-Compensated Avena Seedlings." *Plant Physiology* 43 (1968): 338–344.

[5] Briegleb, W. "Ein Modell zur Schwerelosigkeits-Simulation an Mikroorganismen." *Naturwissenschaften* 54 (1967): 167–167.

[6] Hoson, T., S. Kamisaka, Y. Masuda, M. Yamashita, and B. Buchen. "Evaluation of the Three-Dimensional Clinostat as a Simulator of Weightlessness." *Planta* 203 (1997): S187–S197.

[7] Borst, A.G., and J. van Loon. "Technology and Developments for the Random Positioning Machine, RPM." *Microgravity Science and Technology* 21 (2009): 287–292.

[8] van Loon, J.J.W.A. "Some History and Use of the Random Positioning Machine, RPM, in Gravity Related Research." *Advances in Space Research* 39 (2007): 1161–1165.

[9] Albrecht-Buehler, G. "The Simulation of Microgravity Conditions on the Ground." *ASGSB Bulletin* 5 (1992): 3–10.

[10] John, S.P., and K.H. Hasenstein. "Effects of Mechanostimulation on Gravitropism and Signal Persistence in Flax Roots." *Plant Signaling & Behavior* 6 (2011): 1–6.

[11] Israelsson, D., and A. Johnsson. "A Theory for Circumnutations in Helianthus annuus." *Physiologia Plantarum* 20 (1967): 957–976.

[12] Briegleb, W. "Ein Beitrag zur Frage physiologischer Schwerelosigchkeit." *DVL Cologne* (1967): 7–42.

[13] Schwarz, R.P., T.J. Goodwin, and D.A.Wolf. "Cell Culture for Three-Dimensional Modeling in Rotating-Wall Vessels: An Application of Simulated Microgravity." *Journal of Tissue Culture Methods* 14, no. 2 (1992): 51–57.

[14] Liu, T., X. Li, X. Sun, X. Ma, and Z. Cui. "Analysis on Forces and Movement of Cultivated Particles in a Rotating Wall Vessel Bioreactor." *Biochemical Engineering Journal* 18, no. 2 (2004): 97.

[15] Hammond, T.G., and J.M. Hammond. "Optimized Suspension Culture: The Rotating-Wall Vessel." *American Journal of Physiology* 281(1) (2001): F12.

[16] Nauman, E.A., C.M. Ott, E. Sander, D.L. Tucker, D. Pierson, J.W. Wilson, and C.A. Nickerson. "Novel Quantitative Biosystem for Modeling Physiological Fluid Shear Stress on Cells." *Applied and Environmental Microbiology* 73, no. 3 (2007): 699.

[17] Galie, P.A., D.H. Nguyen, C.K. Choi, D.M. Cohen, P.A. Janmey, and C.S. Chen. "Fluid Shear Stress Threshold Regulates Angiogenic Sprouting." *Proceedings of the National Academy of Sciences of the United States of America* 111, no. 22 (2014): 7968–7973.

[18] Zeng Y., Y. Shen, X.L. Huang, X.J. Liu, and X.H. Liu. "Roles of Mechanical Force and CXCR1/CXCR2 in Shear-Stress-Induced Endothelial Cell Migration." *European Biophysics Journal* 41, no. 1 (2012): 13–25.

[19] Wuest, S.L., S. Richard, I. Walther, R. Furrer, R. Anderegg, J. Sekler, and M. Egli. "A Novel Microgravity Simulator Applicable for Three-Dimensional cell Culturing." *Microgravity Science and Technology* 26 (2014): 77–88.

[20] Kaurov, V. "Lissajous Patterns on a Sphere Surface." http://demonstrations.wolfram.com/LissajousPatternsOnASphereSurface/WolframDemonstrations Project Published: July 1, 2011.

[21] Anken, R.H., U. Baur, and R. Hilbig. "Clinorotation Increases the Growth of Utricular Otoliths of Developing Cichlid Fish." *Microgravity Science and Technology* 22 (2010): 151–154.

[22] Barjaktarovic, Z., A. Nordheim, T. Lamkemeyer, C. Fladerer, J. Madlung, and R. Hampp. "Time-Course of Changes in Amounts of Specific Proteins Upon Exposure to Hyper-g, 2-D Clinorotation, and 3-D Random Positioning of Arabidopsis cell cultures." *Journal of Experimental Botany* 58 (2007): 4357–4363.

[23] De Micco, V., M. Scala, and G. Aronne. "Effects of Simulated Microgravity on Male Gametophyte of Prunus, Pyrus, and Brassica species." *Protoplasma* 228 (2006): 121–126.

[24] Wei, N., C. Tan, B. Qi, Y. Zhang, GX Xu, and H.Q. Zheng. "Changes in Gravitational Forces Induce the Modification of Arabidopsis Thaliana Silique Pedicel Positioning." *Journal of Experimental Botany* 61 (2010): 3875–3884.

[25] Zyablova, N.V., Y.A. Berkovich, A.N. Erokhin, and A.Y. Skripnikov. "The gravitropic and phototropic responses of wheat grown in a space greenhouse prototype with hemispherical planting surface." *Advances in Space Research* 46 (2010): 1273–1279.

15

Vibrations

Daniel A. Beysens[1] and Valentina Shevtsova[2]

[1]CEA-Grenoble and ESPCI-Paris-Tech, Paris, France
[2]Université Libre de Bruxelles, Brussels, Belgium

15.1 Introduction

Vibrating a fluid corresponds to submitting it to a periodic acceleration. We describe below how, in addition to the periodic displacements that result from the vibration, mean movements can follow. Such movements can produce in space the same effects as gravity do on Earth [1] or, alternatively, compensate on Earth the gravity-induced flows as if the fluid were in space. It is this latter aspect that we emphasize in the following.

A vibration can be decomposed into its Fourier harmonic components. For the sake of simplicity, we thus only consider linearly polarized, harmonic vibration whose amplitude X varies with time t as

$$X = a\cos\omega t. \tag{15.1}$$

Here a is amplitude, $\omega = 2\pi f$ is the angular frequency, with f the frequency. When submitted to such a vibration, homogeneous matter is subjected to periodic displacements and acquires periodic velocity $u = -a\omega \sin \omega t$ and acceleration $g = -\omega^2 X$. A fluid, however, is in general not homogeneous in density because it exhibits several phases and/or is involved in mass/heat transfer processes where density gradients are the result of thermal gradients. Likewise, density gradients can be caused by concentration gradients arising from e.g. diffusive process, mixing or rejection/incorporation of solute at the solidifying interface. External accelerations thus act on density gradients and can couple with other gravity-induced flows. Local fluid velocity depends on the local density by inertial effect. It results in local velocity gradients, shear flows and Bernoulli pressure difference, especially across interfaces. Mean displacements, convective flows and instabilities, similar to those induced by buoyancy, can thus follow. They combine with gravity-induced flows and can cancel them.

Phenomena will be different according to the relative importance of the vibration period, $1/f$, and the typical hydrodynamic times, τ (viscous relaxation, thermal diffusion, etc.). What also matters is the relative amplitude vibration with respect to the fluid container size, e. The most interesting situation is the high frequency and small amplitude limit, $\tau f \gg 1$ and $a/e \ll 1$ where local fluid inhomogeneities undergo small vibrations around their mean position while mean flow and interface ordering take place. Typically, $a = 0.1$–2 mm and $f = 1$–100 Hz.

15.2 Thermovibrational Convections

Of particular importance is the effect of vibration on a fluid submitted to a temperature gradient. The latter results in a density gradient sensitive to the vibration. Let us thus consider a fluid submitted to a vibrational acceleration in a thermal gradient in the Rayleigh–Bénard configuration (two parallel plates with a temperature difference ΔT separated by distance e). According to Gershuni and Lubimov [2], vibrational Rayleigh number (ρ is density, p is pressure, T is temperature, D_T is thermal diffusivity) is written as:

$$\mathrm{Rav} = \frac{\left[a\omega\left(\frac{\partial \rho}{\partial T}\right)_p \Delta T e\right]^2}{2\pi D_T}. \tag{15.2}$$

The convection threshold depends on the angle between the thermal gradient and the vibration direction. There is no convection for a temperature gradient parallel to the vibration. The most unstable situation corresponds to a temperature gradient perpendicular to the vibration direction. Here, convection starts when Rav is larger than a few thousands.

15.3 Crystal Growth

These thermovibrational flows can annihilate thermogravitational flow such as buoyancy and/or thermocapillary (Marangoni) convection when an interface is present, depending on the mutual orientation of vibration axis and thermal or compositional gradient. In the system with free interface thermo- (soluto-) capillary (Marangoni) and thermovibrational mechanisms can produce motion in opposite directions. This is particularly the case in a solidification process. Appropriate combination of these mechanisms can then be used to counteract the usual convective flows inherent in crystal growth processes from the liquid phase.

Efficient control of heat and mass transfer during real industrial applications of crystal growth from the liquid phase can be envisaged. The possible utilization of such a strategy for the floating zone crystal growth technique has been addressed [3, 4].

15.4 Dynamic Interface Equilibrium

Harmonic vibrations can considerably deviate the equilibrium position of an interface from its normally horizontal position under gravity acceleration, g, and attain large enough angles [5]. The situation is similar with that of simple mechanical systems under vibration. For example, a simple pendulum of length L can be stabilized in an upside down position by vertically vibrating its support at a frequency much larger than the natural frequency of the pendulum, that is, when $a\omega \geq \sqrt{2gL}$ [6]. When the support of the pendulum is vibrated horizontally at frequencies much larger than the natural frequency of the pendulum and vibrational velocity amplitudes $a\omega$ higher than a threshold value ($a\omega \geq \sqrt{2gL}$), the equilibrium position of the pendulum is no more vertical and makes an angle with respect to the horizontal.

Vibration of a fluid interface can demonstrate similar phenomena. Using vibration, it is possible to stabilize two-fluid configurations, generally unstable when the vibration is absent. For example, a heavier fluid floating over a lighter fluid under terrestrial gravity field is unstable (Rayleigh–Taylor configuration) under normal conditions. Application of strong vertical vibration can dynamically stabilize the above configuration [5]. When subjected to strong horizontal vibration, the interface of an initially horizontal fluid interface can attain a dynamic equilibrium at an angle to the horizontal plane.

Beyond a threshold value of vibrational velocity $a\omega$, the interface attains an equilibrium position at an angle α with vertical (Figure 15.1). The results depend on gravity acceleration g, vibration amplitude and frequency [5]. With L the dimension of the interface the angle can be written as

$$sin\, \alpha = \frac{2gL}{\pi a^2 \omega^2} \frac{\rho_\ell + \rho_v}{\rho_\ell - \rho_v} \tag{15.3}$$

When the density difference between phases is small, the interface can exhibit instability of Kelvin–Helmholtz type called "frozenwave" [7]. It is

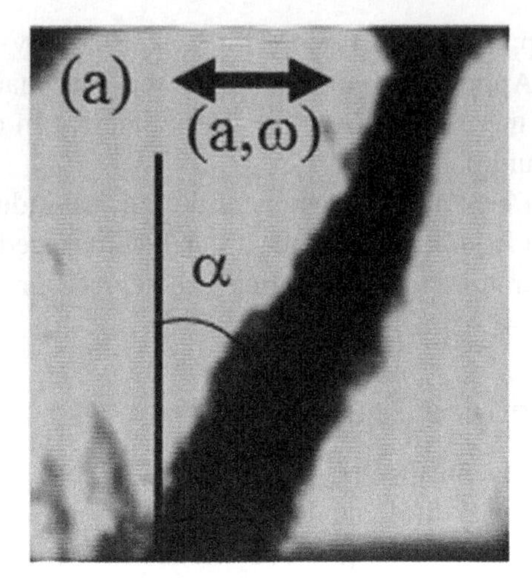

Figure 15.1 Interface position in liquid–vapor hydrogen for the vibration case $a = 0.83$ mm and $f = 35$ Hz and gravity level $0.05g$ (directed vertically). The interface looks fuzzy as it pulsates at the vibration frequency.

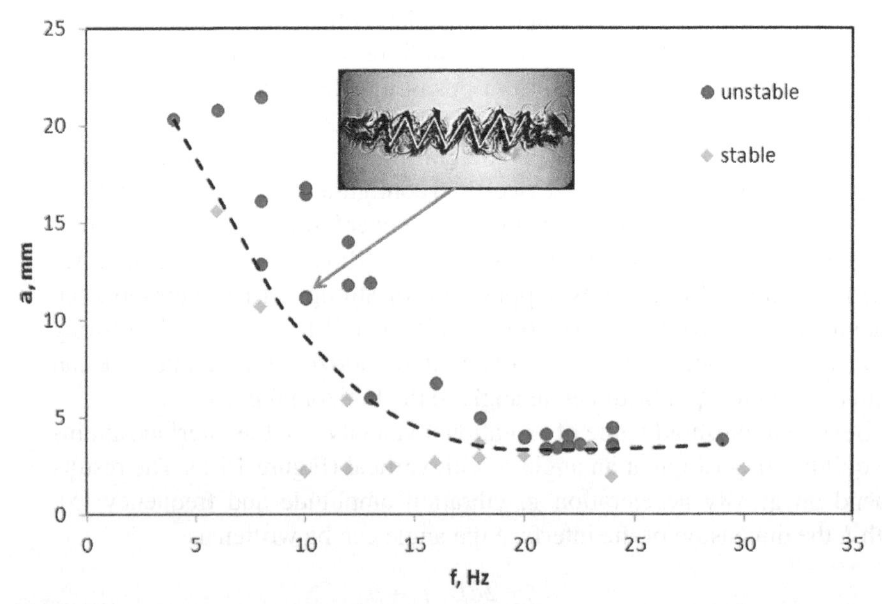

Figure 15.2 Experimental stability map in the plane (a, f) for miscible liquid/liquid interface (mixtures of water–isopropanol of different concentrations). Diamonds: no instability. Circles: instability. The black dashed curve is a guideline for eyes between stable and unstable regions. Inset: Typical shape of the frozen waves with horizontal vibration.

characterized by an interface modulation perpendicular to vibration direction that keeps immobile in the sample reference frame. This instability is also present when two miscible liquids of similar (but nonidentical) viscosities and densities are present [8]. While it was shown experimentally that surface tension can be nonzero between miscible liquids [9], its value is, however, quite small. From the experimental results [8], the values of the critical amplitude were determined as a function of critical frequency as shown in Figure 15.2. In this phenomenon, gravity is in competition with vibrational effect as surface tension does not play a role.

References

[1] Beysens, D. "Critical Point in Space: A Quest for Universality." *Microgravity Science and Technology* 26 (2015): 201–218.

[2] Gershuni, G.Z., and D.V. Lyubimov. *Thermal Vibrational Convection*, New York, NY: Wiley, 1998.

[3] Lyubimov, D.V., T.P. Lyubimova, S. Meradji, and B. Roux. "Vibrational Control of Crystal Growth from Liquid Phase." *Journal of Crystal Growth* 180 (1997): 648–659.

[4] Lappa, M. "Review: Possible Strategies for the Control and Stabilization of Marangoni Flow in Laterally Heated Floating Zones." *Fluid Dynamics & Materials Processing (FDMP)* 1, no. 2 (2005): 171.

[5] Landau, L.D., and E. M. Lifschitz. *Mechanics*. Moscow: Mir, 1973.

[6] Gandikota, G., D. Chatain, T. Lyubimova, and D. Beysens. "Dynamic Equilibrium Under Vibrations of H_2 Liquid-Vapor Interface at Various g-levels." *Physical Review E* 89 (2014): 063003–7.

[7] Gandikota, G., D. Chatain, S. Amiroudine, T. Lyubimova, and D. Beysens. "Frozen Wave Instability in Near Critical Hydrogen Subjected to Horizontal Vibration Under Various Gravity Fields." *Physical Review E* 89 (2014): 012309–1.

[8] Gaponenko, Y.A., M. Torregrosa, V. Yasnou, A. Mialdun, and V. Shevtsova. "Dynamics of the Interface Between Miscible Liquids Subjected to Horizontal Vibrations." *Journal of Fluid Mechanics* (accepted, 2015).

[9] Lacaze, L., P. Guenoun, D. Beysens, M. Delsanti, P. Petitjeans, and P. Kurowski. "Transient Surface Tension in Miscible Liquids." *Physical Review E* 82 (2010): 041606.

4

Other Environmental Parameters

16

Earth Analogues

Inge Loes ten Kate[1] and Louisa J. Preston[2]

[1]Utrecht University, Utrecht, The Netherlands
[2]The Open University, Milton Keynes, UK

Terrestrial analogue environments are places on Earth with geological or environmental conditions that are similar to those that exist on an extraterrestrial body [1]. The purpose of using these terrestrial analogue sites for planetary missions can be divided into four basic categories: (i) to learn about planetary processes on Earth and elsewhere; (ii) to test methodologies, protocols, strategies, and technologies; (iii) to train highly qualified personnel, as well as science and operation teams; and (iv) to engage the public, space agencies, media, and educators [1, 2]. A recent ESA study, CAFE—Concepts for Activities in the Field for Exploration [3], resulted in a catalogue of all planetary analogue sites used and currently in use [4]. This catalogue contains in-depth descriptions of each of these field sites, including location, geological context, environmental information, and infrastructure, and is currently the most extensive and up-to-date catalogue. A very comprehensive overview of analogue sites grouped per planetary surface feature can be found in [5]. Current analogue activities focus on five planetary bodies: the Moon, Mars, Europa and Enceladus, and Titan. Below we highlight a few planetary analogue sites for these five bodies summarized from [5], as well as field-testing campaigns and semipermanent field-testing bases.

16.1 Planetary Analogues

16.1.1 The Moon

The lunar surface features that can be studied in terrestrial analogues are craters, lava fields, and the lunar dust. The Vredefort dome in South Africa was studied as an analogue for the fine-grained granulite facies rocks that were returned from the Moon by the Apollo astronauts [6]. Most lunar analogue sites, however, are chosen to study mission concepts and test instruments. This already started with the Apollo astronaut training in the Lava Mountains, California [7], and the volcanic fields around Flagstaff, Arizona. Specific prerequisites here are aridity, low temperature, and the presence of abrasive dust. Another example of a lunar analogue is the Haughton Impact Structure in Canada [8].

16.1.2 Mars

Most analogue sites are devoted to Mars. These analogue sites can be divided into three categories: early Mars, middle Mars, and present Mars. *Early Mars* is here defined as roughly the first billion years of its lifetime, when liquid water was still presumed to be present on the surface. Example analogue sites are the Pilbara region in Australia as an analogue for flood basalts, water-related minerals, and preservation of early life [9], Rio Tinto in Spain as an analogue for past rivers, iron oxides, and sulfates [10], and Yellowstone as an analogue for silica-rich soils, hydrothermal activity, and extremophiles [11]. *Middle Mars* is defined as the second billion years where a large drop in temperature and loss of water led to a global cryosphere and subsurface ice. The Antarctic Dry Valleys [12] and Antarctic permafrost [13] serve as analogues for the Polar Layered Deposits and the Northern Highlands as well as for potential life preserved in ice deposits. Iceland [14] and the Bockfjord Volcanic Complex on Svalbard, Norway [15], serve as analogues for subglacial volcanism. *Present Mars* starts about 2.5 billion years ago and is characterized by a hyperarid climate. The Antarctic Dry Valleys are a good analogue for present-day Mars and is the closest terrestrial analogue to Mars. Additionally, the Atacama Desert [16], the Egyptian Desert [17], and the Hawaiian volcanoes, for example, Mauna Kea [18], are well-studied Mars analogue sites.

16.1.3 Europa and Enceladus

Both Europa and Enceladus are characterized by a planet-wide ocean covered with a thick ice-crust. Hydrothermal activity at the ocean floor is hypothesized as an energy source to keep the oceans liquid. Analogue sites for all three parts of planets can be found on Earth. Ocean floor analogues can be found in the hydrothermal vents of Lost City on the Mid-Atlantic Ridge [19] and the high-pressure low-temperature environments of the Mariana Trench in the Pacific Ocean [20]. Mono Lake in California, USA, and the Dead Sea in Israel are analogues for the alkaline and saline brine oceans expected on these icy moons. Lake Vostok on Antarctica is one example of a surface ice analogue [21].

16.1.4 Titan

Like the Earth, Titan has a thick atmosphere and diverse geology with land and lakes. These lakes, however, are composed of hydrocarbons, making terrestrial tar fields good analogues for the Titan surface, for example, Pitch Lake in Trinidad and Tobago [22].

16.2 Semipermanent Field-Testing Bases

Long-term field-testing campaigns with some more permanent infrastructure are established to provide a base for multidisciplinary field research as well as for the development of new technologies for planetary missions. Most of these sites are very much technology and mission development focused; however, they do offer the opportunity to carry out scientific campaigns as part of the technology-driven frameworks. These sites include the Aquarius Undersea Research Station at the Florida Keys established in 1993 and host to 114 underwater missions up to 2012, primarily studying coral reefs [23]; the Haughton-Mars Project Research Station at Devon Island, Canada, focusing on "developing new technologies, strategies, and operational protocols geared to support the future exploration of the Moon, Mars, and other planets" [8, 24]; the Pavilion Lake Research Project, a "science and exploration effort to explain the origin of freshwater microbialites in Pavilion Lake, British Columbia, Canada" [25]; the Pacific International Space Center for Exploration Systems (PISCES) at Hawai'i [26, 27]; the Ibn Battuta Centre for exploration and field activities in Morocco, established in 2006 to support the exploration of Mars and others planets, and to provide opportunities for scientists and the public for experiencing the exploration on Earth and in the Solar System [28]; and the recently (2013) established Boulby International Subsurface Astrobiology Laboratory, BISAL, in the UK, the world's first permanent subsurface astrobiology lab, focusing on deep subsurface geochemistry and biology, as well as instrument testing for robotic and human planetary missions [29].

16.3 Field-Testing Campaigns

Long-term field-testing campaigns have been established to provide a framework for planetary instrument and mission testing. The Desert Research and Technology Studies (Desert RATS) field-testing campaigns started in 1997 in support of future manned mission scenarios. The Desert RATS campaigns have taken place in various locations, including Mauna Kea, HI, and Black

Point Lava Flow, AZ. These locations were selected based on their physical resemblance of lunar and martian surface. The scope of these campaigns has varied widely over the years from testing single space suit configurations to multi-day integrated mission scenarios [30]. Even though the focus is mainly on technology-related testing, integrating science into these technology-driven scenarios and getting scientists and engineers to communicate is an important aspect. The NASA Extreme Environments Mission Operations campaigns [31] started in 2001 and 16 missions have been undertaken since then. NEEMO uses the Aquarius Station, since the station habitat and its surroundings provide a convincing analogue for space exploration. Like Desert RATS, NEEMO is rather technology and mission oriented. The Arctic Mars Analog Svalbard Expedition (AMASE) at Svalbard, Norway [32], is an astrobiology- and Mars-focused science and technology campaign taking place on Svalbard, Norway. This campaign is specifically focused on the understanding of Svalbard in an astrobiological context. Technology that is taken along is tested in support of the science and not the other way around, as is merely the case in, for example, Desert RATS and NEEMO.

References

[1] Léveillé, R. "Validation of Astrobiology Technologies and Instrument Operations in Terrestrial Analogue Environments." *Comptes Rendus Palevol* 8, no. 7 (2009): 637–648.

[2] Lee, P. "Haughton-Mars Project 1997–2007: A Decade of Mars Analog Science and Exploration Research at Haughton Crater, Devon Island, High Arctic." In *Proceedings of 2nd International Workshop on Exploring Mars and its Earth Analogs*. Pescara, Italy: International Research School of Planetary Sciences (IRSPS), 2007.

[3] Preston, L.J., S. Barber and M. Grady. *CAFE—Concepts for Activities in the Field for Exploration—Executive Summary Report*. ESA Contract # 4000104716/11/NL/AF, 2013. Last visited on November 15, 2013. http://esamultimedia.esa.int/docs/gsp/C4000104716ExS.pdf.

[4] Preston, L.J., S. Barber and M. Grady. *CAFE—Concepts for Activities in the Field for Exploration—TN2: The Catalogue of Planetary Analogues*, 2013. Last visited on November 15, 2013. http://esamultimedia.esa.int/docs/gsp/The_Catalogue_of_Planetary_Analogues.pdf.

[5] Preston, L.J. and L.R. Dartnell. "Planetary Habitability: Lessons Learned from Terrestrial Analogues". *International Journal of Astrobiology* 13,no. 01 (2014): 81–98.

[6] Gibson, R.L., W.U. Reimold, A.J. Ashley and C. Koeberl. "Metamorphism on the Moon: A Terrestrial Analog in the Vredefort Dome, South Africa?" *Geology* 30, no. 5 (2002): 475–478.

[7] Hinze, W.J., R. Ehrlich, H.F. Bennett, D. Pletcher, E. Zaitzef and O.L. Tiffany. "Use of an Earth Analog in Lunar Mission Planning". *Icarus* 6, no. 1–3 (1967): 444–452.

[8] Osinski, G.R., P. Lee, C.S. Cockell, K. Snook, D.S.S. Lim and S. Braham. "Field Geology on the Moon: Some Lessons Learned from the Exploration of the Haughton Impact Structure, Devon Island, Canadian High Arctic. *Planetary and Space Science* 58, no. 4 (2010): 646–657.

[9] Allwood, A.C., M.R. Walter, I.W. Burch and B.S. Kamber. "3.43 Billion-Year-Old Stromatolites Reef from the Pilbara Craton of Western Australia: Ecosystem-Scale Insights to Early Life on Earth". *Precambrian Research* 158 (2007): 198–227.

[10] Amils, R., E. González-Toril, D. Fernández-Remolar, F. Gómez, Á. Aguilera, N. Rodríguez, M. Malki, A. García-Moyano, A.G. Fairen, V. de la Fuente and J. Luis Sanz. "Extreme Environments as Mars Terrestrial Analogs: The Rio Tinto Case". *Planetary and Space Science* 55, no. 3(2007): 370–381.

[11] Barns, S.M., R.E. Fundyga, M.W. Jeffries and N.R. Pace. "Remarkable Archaeal Diversity Detected in a Yellowstone National Park Hot Spring Environment". *PNAS* 91 (1994): 1609–1613.

[12] Wentworth, S.J., E.K. Gibson, M.A. Velbel and D.S. McKay. "Antarctic Dry Valleys and Indigenous Weathering in Mars Meteorites: Implications for Water and Life on Mars". *Icarus* 174, no. 2 (2005): 383–395.

[13] Dickinson, W.W. and M.R. Rosen. "Antarctic Permafrost: An Analog for Water and Diagenetic Minerals on Mars". *Geology* 31, no. 3 (2003): 199–202.

[14] Cousins, C.R. and I.A. Crawford. "Volcano–Ice Interaction as a Microbial Habitat on Earth and Mars". *Astrobiology* 11 (2011): 695–710.

[15] Treiman, A.H., H.E.F. Amundsen, D.F. Blake and T. Bunch. "Hydrothermal Origin for Carbonate Globules in Martian Meteorite ALH84001: a Terrestrial Analogue from Spitsbergen (Norway)". *Earth and Planetary Science Letters* 204 (2002): 323–332.

[16] Navarro-Gonzàlez, R., et al. "Mars-Like Soils in the Atacama Desert, Chile, and the Dry Limit of Microbial Life." *Science* 302, no. 5647 (2003): 1018–1021.

[17] Heggy, E. and P. Paillou. "Probing Structural Elements of Small Buried Craters Using Ground-Penetrating Radar in the Southwestern Egyptian

Desert: Implications for Mars Shallow Sounding". *Geophysical Research Letters* 33, no. 5 (2006): L05202.

[18] ten Kate, I.L., R. Armstrong, B. Bernhardt, M. Blumers, J. Craft, D. Boucher, E. Caillibot, J. Captain, G.M.T. D'Eleuterio, J.D. Farmer, D.P. Glavin, T. Graff, J.C. Hamilton, G. Klingelhöfer, R.V. Morris, J.I. Nuñez, J.W. Quinn, G.B. Sanders, R.G. Sellar, L. Sigurdson, R. Taylor and K. Zacny. "Mauna Kea, Hawai'i, as an Analogue Site for Future Planetary Resource Exploration: Results from the 2010 ILSO-ISRU Field-Testing Campaign". *Journal of Aerospace Engineering* 26, no. 1(2013): 183–196.

[19] Kelley, D.S., et al. "An Off-Axis Hydrothermal Vent Field Discovered Near the Mid-Atlantic Ridge at 30°N". *Nature* 412(2001): 145–149.

[20] Sharma, A., J.H. Scott, G.D. Cody, M.L. Fogel, R.M. Hazen, R.J. Hemley and W.T. Huntress. "Microbial Activity at Gigapascal Pressures". *Science* 295 (2002): 1514–1516.

[21] Ellis-Evans, J.C. and D. Wynn-Williams. "A Great Lake Under the Ice". *Nature* 381 (1996): 644–646.

[22] Meckenstock, R.U., et al. "Water Droplets in Oil are Microhabitats for Microbial Life". *Science* 345, no. 6197(2014): 673–676.

[23] Todd, B. and M. Reagan. "The NEEMO Project: A Report on How NASA Utilizes the "Aquarius" Undersea Habitat as an Analog for Long-Duration Space Flight". *Engineering, Construction, and Operations in Challenging Environments*(2004): 751–758.

[24] Lee, P. and G.R. Osinski. "The Haughton-Mars Project: Overview of Science Investigations at the Haughton Impact Structure and Surrounding Terrains, and Relevance to Planetary Studies". *Meteoritics and Planetary Science* 40, no. 12(2005): 1755–1758.

[25] Lim, D.S., A.L. Brady and Pavilion Lake Research Project (PLRP) Team. "A Historical Overview of the Pavilion Lake Research Project—Analog Science and Exploration in an Underwater Environment". Special Paper 483. Boulder, CO: Geological Society of America, 2011, 85–116.

[26] Schowengerdt, F., R. Fox, M. Duke, N. Marzwell and B. McKnight. "PISCES: Developing Technologies for Sustained Human Presence on the Moon and Mars." In *Proceedings, 3rd AIAA Space Conference and Ex-position*. Reston, VA: American Institute of Aeronautics and Astronautics (AIAA), 2007, 3029–3038.

[27] Duke, M.B., et al. "PISCES: Hawaii Facility for Simulation and Training." In Proceedings of 38th Lunar and Planetary Science Conferency. Houston: Lunar and Planetary Institute (LPI), 2007.

[28] Cavalazzi, B., R. Barbieri and G.G. Ori. "Chemosynthetic Microbialites in the Devonian Carbonate Mounds of Hamar Laghdad (Anti-Atlas, Morocco)". *Sedimentary Geology* 200 (2007): 73–88.

[29] www-1: http://www.astrobiology.ac.uk/research/bisal/. Last visited on December 5, 2014.

[30] Ross, A., J. Kosmo and B. Janoiko. "Historical Synopses of Desert RATS 1997–2010 and a Preview of Desert RATS 2011". *Acta Astronautica* 90, no. 2(2013): 182–202.

[31] Thirsk, R., D. Williams and M. Anvari. "NEEMO 7 Undersea Mission". *Acta Astronautica* 60, no. 4–7(2007): 512–517.

[32] Steele, A., H.E.F. Amundsen and AMASE 07 Team. "Arctic Mars Analog Svalbard Expedition 2007". In *38th Lunar and Planetary Science Conference*. Houston: Lunar and Planetary Institute (LPI), 2007.

17

Isolated and Confined Environments

Carole Tafforin

Ethospace, Toulouse, France

Characteristics of space analogue environments with regard to human performance concern the crew adaptation in a socio-psychological context and in a temporal dynamics. Isolation, confinement and time are major features on Earth to reproduce an extra-terrestrial environment for manned mission simulations. In the current space missions (low Earth orbit, LEO) and in the perspective of interplanetary missions (near-Earth asteroid, Moon, Mars), men and women will have to adapt to social constraints (crew size, multinationality, mixed-gender) and spatial restrictions (volume, multi-chambers, life-support) on short-term, medium-term and long-term durations. The crewmembers also will have to perform intra-vehicular activities (IVA) and extra-vehicular activities (EVA). For training, preventing and optimizing such tasks, simulations of living and working together in isolated and confined environments, and simulations of operating with a space suit on geological surfaces are the new requirements.

During the two decades (1991–2011), space simulators (confinement) and analogue settings (isolation) were adequately developed on Earth with the ultimate goal of walking on Mars. Time periods extended up to 500 days. Space analogue environments are located worldwide (Canada, United States, Russia, Europe, Antarctica and Arctic). Mission durations in space analogue environments cover days, weeks and years. Isolation and confinement facilities implemented for such simulations are listed in Table 17.1.

Over a 7-day duration, the Canadian Astronaut Program Space Unit Life Simulation (CAPSULS) was an Earth-based initiative that simulated a typical space shuttle or space station mission [1]. CAPSULS provided the Canadian astronaut participants with space mission training. The facility was a solid

Table 17.1 Isolation and confinement facilities implemented for such simulations

Name	Type	Site	Year	Crew Size	Nationality	Gender	Volume	Duration
ISEMSI	Multi-chamber	Bergen, Norway	1990	6	European	Males	118 m^3	28 days
EXEMSI	Multi-chamber	Cologne, Germany	1992	4	European	Males	95 m^3	60 days
HUBES	Multi-chamber	Moscow Russia	1994	3	Russian	Males	100 m^3	135 days
BIOSPHERE 2	Desert station	Arizona, USA	1991	8	Multi-	Mixed	1.27 ha	730 days
			1994	7	Multi-	Mixed	(surface)	6 months
CAPSULS	Hyperbaric chamber	Toronto Canada	1994	4	Canadian	Mixed	33 m^3	7 days
SFINCSS	Multi-chamber	Moscow Russia	1999–2000	33×2	Russian Multi-	Males Mixed	100 m^3 200 m^3	240 days 110 days
FMARS	Polar station	Devon Is. Canada	2000–2007	6–7	Multi-	Mixed	416 m^3	7 days-4 months
NEEMO	Undersea missions	Florida, USA	2001–2004	6–9	American	Mixed	401 m^3	6–12 days
MDRS	Desert station	Utah, USA	2002	6–8	Multi-	Mixed	500 m^3	15 days
CONCORDIA	Polar station	Dome C, Antarctic	2006–	10	French, Italian	Mixed	1,500 m^2	8 months
TARA	Polar ship	Ice pack, Arctic	2008	10	Multi-	Mixed	36 m	507 days
MARS-500	Multi-chamber	Moscow, Russia	2010–2011	6	Multi-	Males	550 m^3	520 days

steel hyperbaric chamber that was comprised of two primary modules, the habitat module and the experiment module, which were connected by a smaller transfer chamber. During the simulation, crew inside monitored and controlled life-support parameters (temperature, humidity, pressure of oxygen, carbon dioxide, nitrogen and water vapor). CAPSULS involved national (Canadian) mixed-gender and small-size crew for short-term simulations. They emphasized social and personality issues of adaptation.

Over a 6- to 12-day duration, NASA Extreme Environment Mission Operations (NEEMO) project was an analogue mission that sent groups of astronauts, engineers and scientists to live in Aquarius, the world's only undersea research station [2]. The Aquarius habitat and its surroundings provide a hostile environment with the risk of decompression sickness in depth immersion. NEEMO crewmembers, named Aquanauts, experienced some of the same challenges that they would on a distant asteroid or on Moon. During NEEMO missions, the aquanauts simulated living on a spacecraft and tested spacewalk techniques. Crew size was equivalent to a real space crew. Like CAPSULS, there were mainly national (USA) but mixed-gender crew and for short-term simulations. Underwater condition had the additional benefit to simulate a microgravity environment by buoyancy. The peculiarity

Figure 17.1 Tara expedition in Arctic (Image © F. Latreille/Tara expédition).

of NEEMO is that crewmembers participated in undersea sessions of EVA and used autonomous life-support systems during IVA (Kanas et al. 2010).

Over a 15-day duration, Mars Research Desert Station (MDRS) was one of four stations planned by the International Mars Society [3] which has established a number of prototype Mars Habitat Units around the world. MDRS is a 2-deck facility providing a comprehensive living and working environment for a Mars or Moon crew. The upper deck includes sleeping quarters, a communal living area, a small galley, exercise area and hygiene facilities with closed-circle water purification. The lower deck includes the primary working space for the crew: small laboratory areas for carrying out geology and life science research, storage space for samples, airlocks for reaching the surface of planets, and a suiting-up area where crewmembers prepare for surface operations. MDRS also involved small-size but multinational and mixed-gender crews, for longer-term simulations. They thus emphasized cultural issues. The specificity is EVA while walking in a desert landscape mimicking a Mars landscape.

Over a 28-day duration, Isolation Study for the European Manned Space Infrastructure (ISEMSI) was one of three European campaigns implemented as precursor flights regarding medium-term orbital stays [4]. ISEMSI was a confined habitat in multi-chamber facilities composed of 6 modules. Two small modules served as sleeping chambers, the larger module served as working and living chamber, one module was used as sanitary, one module acted as a transfer lock and storage and one module was a rescue chamber fully equipped to maintain life-support functions under pressure. As in NEEMO and MDRS, the crew size was as expected for space mission and space exploration but

Figure 17.2 Concordia station in Antarctica (Image © IPEV).

with only male crewmembers. ISEMSI campaign was specifically designed for psychological and physiological issues on long-duration manned space mission.

Over a 60-day duration, Experimental Campaign for the European Manned Space Infrastructure (EXEMSI) prolonged ISEMSI campaign, thus doubling the time spent in a confined habitat. The overall facility consisted of two main modules serving respectively as the habitat and the laboratory module for carrying out scientific investigations and teleoperations, completed with two transfer and storage modules [5]. The living quarters containing bunks and a table were used for sleeping, eating and leisure time. In the panel of confined and isolated settings, ISEMSI had the smallest volume and the smallest crew size. It differed from NEEMO and MDRS because it provided only IVA but promoted further advances on long-duration manned space mission.

Over a 135-day duration, Human Behavior in Extended Spaceflight (HUBES) prolonged ISEMSI and EXEMSI campaigns and was definitely considered as a space simulator for extended periods [6]. The confined habitat was designed as a unique module, thus sharing living area, working area and storage area in a small volume compared to CAPSULS. HUBES had the benefit to reproduce real mission aboard the Mir orbital station and crew transfer flight considering the very small crew size consisting of Russian and male-only crewmembers. HUBES was a ground-based simulation aimed at comparing and validating psychological issues in crew selection, training, monitoring and in-orbit support flight.

Over a 4-month duration, Flashline Mars Arctic Research Station (FMARS) hosted a long-duration mission that quadrupled the in situ simulation of both confinement and isolation. Its geographic location, the polar

Figure 17.3 Mars Desert Research Station in Utah/USA (Image © Ethospace).

desert, is the largest uninhabited island in the world [7]. The 24-h daylight is analogous to that of a polar Moon or Mars mission. FMARS and MDRS have the same Mars Habitat Unit. The difference is the simulation of Mars-Earth latency to a 20-min delay and the tightly life-support since water use was restricted and monitored. However, the facility was not hermetically locked as enforced during NEEMO mission. Important analogue features of FMARS were EVAs and the hostile surroundings in Arctic. The multinational and mixed-gender composition of the crew properly simulated future isolated and confined crews who would land on asteroids, Moon or Mars.

Over a 110 to 240-day duration, Simulation for Flight of International Crew on Space Station (SFINCSS) was devoted to determine the extent to which monotony may impair individual performance, physical rate and transcultural crew performance. The facilities consisted of either one main chamber or two chambers; separated groups lived on a schedule typical for Mir station crews then on a fixed work–rest schedule mimicking International Space Station (ISS) assembly operations [8]. Specific feature of SFINCSS was the participation of several small-size and multinational crews. Lessons learned from ISEMSI, EXEMSI, HUBES and SFINCSS provided an overview of issues involved in the design and habitability of simulators and simulation campaigns.

Figure 17.4 Mars-500 experiment in Moscow/Russia (Image © Ethospace).

Over an 8-month duration, Concordia South Pole station hosted winterers in high isolation and confinement conditions approaching those found in very long-term interplanetary flights. The facility consisted of two cylindrical structures designed as a space architecture [9]. Antarctica is the farthest and most hostile environment on Earth because of the lowest outside temperatures, the windy and sunny conditions that require the crew being strictly isolated and confined for safety reasons. The inner volume of habitat is very large compared to all other space simulators. Concordia station is furnished with a life-support system analogous to that implemented onboard the European Columbus module in the ISS. Such extreme environment addresses socio-psychological and medical issues of a relatively large-size and binational (French and Italian) crew.

Over a 507-day duration, Tara expedition in Arctic was a polar schooner that once embedded in the ice, drifted on the pack until reaching the free oceanic waters [10]. Tara was not originally devoted to space researches but the drift was analogous to a flight to Mars and return to Earth. The very small volume of habitat and the very hostile surrounding in North Pole were actually relevant features of a confined and isolated environment. The crewmembers, named Taranauts, experienced continuous polar night (the sun disappeared for more than 4 months) and continuous polar day for more than 3 months, like at Concordia. The large-size crew was multinational and mixed-gender and composed of scientists, technicians, medical doctor, artist, photographer, sailor and diver. The heterogeneity of the crew was specific to Tara expedition. The dangerous nature of such mission in real conditions addressed stress and coping issues.

Over a 520-day duration, Mars-500 experiment was the longest designed analogue Mars mission of the world [11]. This experiment was conducted entirely in confinement scheduled on a 250-day interplanetary flight from Earth to Mars, a 30-day Mars orbital stay and surface operations, and a 240-day interplanetary flight from Mars to Earth. Furthermore, a 20-minute delay communication was simulated with unexpected tests of loss of communication. Mars-500 facilities were composed of four hermetically sealed chambers for ensuring life-support functions: the habitable module, the medical module, the storage module and the Mars-landing module. They were connected to an external module: Martian surface simulator. The Marsionauts were only males but involved Russian, European and Chinese cultural backgrounds. They were separated into orbital crew and landing crew, the latter one simulating EVAs while walking on the Mars surface. Mars-500 provided a unique insight into human factor and cultural factor issues for future explorers of distant planets. The emphasis was on crew autonomy.

Over a 730-day duration, Biosphere 2 was a concept similar to the housing structure on extra-terrestrial environment. It was the longest experience in the field. The facilities were built to mainly be an artificial, materially closed ecological system and to explore human colonization of other planets. The wide landscape integrated fauna, flora, mountain area, desert area, water area and any atmospheric parameters in order for the crew to survive. The crew was specifically composed on a male/female quota. Biosphere 2 project was a full-scale life-support model and the adaptation of crews to isolated and confined environment beyond two years opened new issues for creating new ground-based simulations devoted to space science.

Other older settings about polar missions at Vostok and Mac Murdo, for instance, are permanent Antarctic stations which particularly emphasized the isolation parameter whereas submarine missions were relevant to the confinement parameter. Designing and building environments for humans in outer space is the main challenge to ensure and study isolated and confined crews' behavioral health [12]. New programs are being launched today. Several 7-day missions in the Human Exploration Research Analog (HERA) habitat at NASA Johnson Space Center in Houston, Texas, are planned to emulate what might be required on a mission to an asteroid. The current Hawaii Space Exploration Analog and Simulation (HI-SEAS) in a sealed Dome near Mauna Loa volcano is conducted for the next 8 months, to train for what conditions would be like on Mars.

In the future, the replication of such protocols would be needed for salient scientific outcomes. We could consider a continuous study, on a permanent site, within a unique facility, with a crew change every 500 days.

Acknowledgement

Special thanks to Pr. Christian Tamponnet for his support.

References

[1] Canadian Space Agency–CAPSULS. www.asc-csa.gc.ca/eng/astronauts/osm_capsuls.asp
[2] NASA–NEEEMO. www.nasa.gov/mission_pages/NEEMO/index.html
[3] Mars Desert Research Station–The Mars Society mdrs.marssociety.org

[4] Collet, Jacques. "The first European simulation of a long-duration manned space mission." In *Advances in Space Biology and Medicine*, edited by Sjoerd L. Bonting. Greenwich London: JAI Press Inc., 1993.

[5] Vœrnes, Ragnar J. "EXEMSI: descriptive of facilities, organization, crew selection, and operational aspects." In *Advances in Space Biology and Medicine*, edited by Sjoerd L. Bonting. Greenwich London: JAI Press Inc., 1996.

[6] Tafforin, Carole, and Alessandro Bichi. "Global Analysis of Scientific Data from the three experimental campaigns for European Manned Space Infrastructure." In *Proceeding of the 26th International Conference on Environmental Systems.* Monterey, California, United States. SAE International # 961442, July 1993.

[7] Flashline Mars Artic Research Station–The Mars Society. http://www.fmars.marssociety.org

[8] Mohanty, Susmita, Sue Fairburn, Babara Imhof, Stephen Ransom, and Andreas Volger. "Survey of past, present and planned human space mission simulators." In *Proceeding of the 38th International Conference on Environmental Systems.* San Francisco, California, United States. SAE International # 2008-01-2020, June 2008.

[9] Institut Paul Emile Victor–Concordia. http://www.institut-polaire.fr/ipev/bases_et_navires/station_concordia_dome_c

[10] Home–Tara. http://arctic.taraexpeditions.org

[11] European Space Agency–Mars500. www.esa.int/Our_Activities/Human_Spaceflight/Mars500

[12] Harrison, Albert A. "Humanizing Outer Space: Architecture, Habitability, and Behavioral Health." *Acta Astronautica* 66 (2010): 890–896.

5

Current Research in Physical Sciences

18

Fundamental Physics

Greg Morfill

Max Planck Institute for Extraterrestrial Physics, Garching, Germany

18.1 Introduction

In the late 1800s, physics appeared too much as a field that had reached its limit. It was thought that given enough perseverance everything could be understood, based on established laws like electromagnetics, mechanics and hydrodynamics.

Then came a revolution—new fields, for example, quantum mechanics, relativity, elementary particles and later deterministic chaos, nanophysics, changed the way the world was perceived.

As is so often (practically always in research) the case, these fundamental discoveries raised more questions than they could answer, and new branches of research emerged.

Now after about 100 years, there is a conviction that the next "big frontier" is Biology—understanding cells, cell interactions, the genome, proteins, enzymes, etc. There is clearly a lot of truth in this and physics has played an important part in making this possible by developing the fundamentals of electron microscopy, cryotomography, etc.—the tools needed to make advances in biology possible.

But is physics at its fundamental level really understood today? Has the discovery of the Higgs Boson tied up the missing link in elementary particle physics and field theory? Are there serious questions still unanswered or is physics now entering an era of incremental advances with no major breakthrough to be expected?

In this section, we will summarize some of these outstanding major issues, the approaches made towards tackling these and, in particular, the role played by microgravity and space research.

18.2 The Topics

Some of the most intriguing issues in fundamental physics today are the following:

1. On cosmic scales
   - The incompatibility of quantum mechanics and general relativity
   - Quantum entanglement and action at a distance
   - The nature of "dark matter" and "dark energy"
   - Constancy of "fundamental constants" in time and space
   - Compatibility of inertial and gravitational mass
   - Physics inside black holes
   - Origin of the universe—quantum gravity
2. On more everyday scales
   - Nature of the onset of cooperative phenomena
   - Atomistic understanding of fluids
   - Origin and onset of instabilities
   - Origin and onset of turbulence
   - Limits of hydrodynamics—transition to nanofluidics
   - Phase transitions (equilibrium/non-equilibrium)
   - Critical point phenomena
   - Renormalization group theory at the particle level

In other words, the behavior and self-organization of matter at the individual particle level appears to be one of the most interesting regions of contemporary research. This is driven not only by the pure quest of knowledge, but there are also application interests involved as systems get smaller and smaller.

Where does "Space" and in particular "microgravity" enter in our quest to learn more about these fundamental issues? And why is it important to probe these topics further?

To provide an answer to the second question is to look at history. Without quantum mechanics, we would not have semiconductors, computers of a quality provided by the smart phone processors would occupy an indoor basketball court, most medical diagnostic and therapeutic devices (e.g., tomography, pacemakers and hearing aids) would not exist—too large and prohibitively expensive—and many other devices we have become used to, like satellite navigation (which would be useless within hours without knowledge of general relativity), lasers, smart phones and telecommunication, would still be science fiction. So to conclude, every major physics breakthrough has the potential at some point in the future to become utilized for the benefit of humanity (but also for our detriment, if used wrongly). But this is common wisdom, and the ethics of utilizing or refraining from utilizing knowledge is always going to be with us—whether it is throwing a stone, deploying an atomic bomb or manipulating the genome. Whatever it is, if enough research is done to understand the consequences, then well-founded decisions are possible. If the understanding is faulty, decisions are very likely faulty, too.

So basic research is necessary and (mostly) beneficial if used correctly.

The answer to the second question, "Where does space research play an important or even a decisive role in fundamental physics?" will be summarized next with a few examples. There is not enough scope in this chapter for a complete treatment of all the interesting topics, so some selection has to be made.

18.3 Fundamental Physics in Space

We have listed some of the major outstanding questions in fundamental physics above. Now the utilization of the special conditions offered by space has to be evaluated. Experiments in space are difficult and costly, so the return (in terms of knowledge) has to be correspondingly great. Otherwise the economic reality will soon overcome the scientists' dreams.

- Space offers a world without gravity. Since gravity is one of the fundamental topics, this fact alone makes space a very attractive proposition for new and novel experiments that cannot be performed on Earth.
- Space is huge. Distance is of great relevance in many fundamental experiments, so here again space provides an attractive environment.
- Space is undisturbed. On Earth many environmental effects "contaminate" measurements, especially as the precision gets increasingly important.

It is not surprising, therefore, that proposals for utilizing the unique space environment have received a strong support throughout the scientific community. Highly rated fundamental research topics are as follows:

- Quantum communication
- Wave-particle duality
- Quantum gases/Bose–Einstein condensation (BEC)
- Atom interferometry
- Constancy of fundamental constants
- Critical point studies in colloids and complex plasmas

- Solidification of colloids in space: Structure and dynamics of crystal, gel and glassy phases

Such a broad and technologically novel approach using the special conditions for research offered by space—promising giant steps in our understanding of physics—was last seen at the beginning of the twentieth century. Today the "enabling factor" is the availability of research under microgravity conditions, in particular the ISS.

One of the major puzzles—perhaps even the major puzzle in physics—is the *incompatibility between "General Relativity Theory"* and *"Quantum Theory"*. Both theories have been tested and verified to typically 1 part in 10^{10} quantitatively, and must be regarded as very sound. Nevertheless, they are incompatible. A great deal of research effort is spent to understand this, but so far no convincing explanation is forthcoming. One possible resolution of this puzzle is self-gravity, which could destroy the particle wave function. Experiments to test this (e.g., massive particle interferometry, massive BEC interactions) need microgravity.

Another major question concerns the *fundamental constants*, for example, the gravitational constant, the fine structure constant, Planck's constant, the elementary charge, the proton/electron mass and speed of light. Are these "constants" really universally constant, or do they vary with time on time scales (and accordingly length scales) of the age of the universe? A possibility to test this requires enormously precise and stable clocks. These are usually based on atomic or optical processes. Comparisons can provide new thresholds of constancy or perhaps even measure possible time effects. Stable clocks require microgravity.

Then there is the issue of *"gravitational mass"* and *"inertial mass"*, as discussed in the famous "equivalence principle." Are they really the same? And how precisely can we measure the predicted gravitational redshift? Such experiments can only be conducted in space if we wish to push the limits of detection to new records.

And last but not least, there is the topic of *"mesoscopic quantum states" and the issue of the "wave–particle duality"*. On the one hand, this concerns Bose–Einstein condensates of comparatively huge (billions of elementary) masses, the interactions between such mesoscopic quantum states and the possible effect of self-gravitation and quantum entanglement. Such massive BECs require microgravity in order to grow (and cool) them. On Earth, they cannot be trapped long enough. On the other hand, one would like to investigate wave properties of large particles using modern versions of the "double-slit" experiment. Since the de Broglie wavelength of particles is inversely proportional to their mass and proportional to their velocity, going to larger particles requires lower velocities (and consequently longer time scales while the particles are moving)—an impossible constraint to maintain on Earth under gravity.

18.3.1 Fundamental Issues in Soft Matter and Granular Physics

"Soft matter" is a name given by the 1991 Nobel Prize Laureate Pierre-Gilles de Gennes to a class of substances (e.g., polymers, colloids, gels and foams) that exhibit macroscopic softness and whose structure and dynamics is not governed by quantum effects (e.g., mesoscopic and supramolecular materials and material assemblies). "*Soft matter*" describes a broad interdisciplinary field covering physics, chemistry and biology, with applications as disparate as paints, new and extreme materials, functionalized (bio) surfaces, etc. Two "recent additions" to this field of soft matter are "complex plasmas" and "granular matter".

The need for experiments in space again stems from the gravity-free environmental conditions. Under microgravity, some systems are easier to produce, and fragile structures can survive longer. Processes such as convection are absent and therefore cannot inhibit delicate structure formation. Finally, there is the topic of self-organization and dynamical processes at the atomistic level.

Experiments in complex plasma physics have been conducted on the ISS since the very beginning, a period covering 14 years so far. During this long time, the research focus has evolved considerably.

In the early years, the emphasis was on researching the properties of this "new state of matter"—the structure of plasma crystals, propagation of waves, domain boundaries, dislocations, crystallization fronts, melting, etc.—all at the "*atomistic*" level of the motion of individually resolved interacting microparticles, with a temporal resolution fine enough to investigate the dynamics all the way into the range of, for example, the Einstein frequencies in crystals, thus providing access to a physical regime that was previously not accessible for studies at this level. In the last few years, it has been realized that "active" experiments can provide an even bigger and more ambitious scope. The focus now includes the following (remember all studies at the most basic "*atomistic*" level):

- Fundamental stability principles governing fluid and solid phases.
- Non-equilibrium phase transitions (e.g., electro-rheology).

- Phase separation of binary liquids.
- The principles of matter self-organization.
- Universality concepts at the kinetic level in connection with critical phenomena (with the long-term aim of understanding the kinetic origin of renormalization group theory, as developed by a Nobel Laureate Kenneth Wilson)
- The physics (structure and dynamics) on approaching the onset of cooperative phenomena in "small" nano-systems.
- The kinetic origin of turbulence.
- Non-Newtonian physics effects.

So far it has been demonstrated that complex plasmas—with their unique properties of visualization of individual particles and comparatively slow (10^{-2} s) dynamic time scales—can contribute enormously to all these areas of research. On Earth, these studies are complemented by two-dimensional systems since gravity forces acting on the (comparatively heavy) microparticles lead to flat membrane-like assemblies. Two-dimensional studies are of great interest too, so that this complementarity is very valuable. The tasks ahead are to utilize existing and new laboratories on the ISS for dedicated experiments to study these basic strong coupling phenomena and to link the observations to the complementary 3D research carried out in complex fluid studies. The two fields—complex plasmas and complex fluids—may be thought of as different states of soft matter (relating to the "gaseous" or "plasma" state and "liquids" respectively) with correspondingly different properties.

In most of the research topics in dust physics, microgravity provides a unique even essential environment. For one thing, *interstellar, protostellar and planetary ring dust phenomena occur under weightlessness* (so it appears reasonable to also use such conditions in experiments) but in addition, some processes require adequate observation time and controlled environments that cannot be achieved in the Earth's gravitational field.

At first glance, it seems strange for "*granular matter*"—*close packed assemblies of near-identical and/or size distributed particles*—to be researched in space under microgravity conditions, especially when vibrations are employed to create an artificial gravity. What is mostly not realized is the enormous scope of granular matter in industry (sand, gravel, grains, etc. are the most obvious, and on the finer scale are toner particles, colloids, paints, etc.) and the surprisingly complex issues involved in size sorting, storage, stability, transport, filling, etc. Size sorting can be achieved under gravity by, for example, vibration, and the larger particles then migrate to the surface—somewhat counterintuitive since they are heavier.

In order to study the processes involved in granular matter physics to understand and model them for the benefit of better and more controlled application on Earth, it is imperative to vary the parameters influencing these processes. One of these parameters, on Earth a constant, is gravity. In space, gravity is absent (or very small). This has several benefits for fundamental studies:

- bigger particles can be used
- time scales for experiments are larger
- the role of fluctuating forces (e.g., vibration) can be studied without "interference" by a macroscopic directed force
- processes can be studied under controlled and variable conditions
- reliable models can be developed that can benefit industrial processes

While all of this seems very "application oriented," there is also a fundamental aspect to this research. This has to do with the self-organization of "hard sphere" matter. In complex plasmas and complex fluids, we discussed strongly interacting systems with a soft interaction potential (a Debye–Hückel potential in the case of complex plasmas) on the one hand and an overdamped hard sphere potential (complex fluids) on the other hand. Granular matter closes a "systemic gap" by providing a virtually undamped hard sphere system. In this sense, a new regime of parameter space becomes available for studying self-organization processes.

References

Fundamental Interactions—Quantum Physics in Space Time

[1] *Connecting Quarks with the Cosmos: Eleven Science Questions for the New Century*. Board on Physics and Astronomy, The National Academies of Press, 2003.

[2] *Assessment of Directions in Microgravity and Physical Sciences Research at NASA*. Space Studies Board, The National Academies of Press, 2003.

[3] European Physics Society (EPS) Position Paper. *The Need for Space Flight Opportunities in Fundamental Physics*. EPS, 2005.

[4] Schutz, B.F. *Fundamental Physics in ESA's Cosmic Visions Pan(PDF)*. ESA Publication SP-588, 2005.

[5] *Revealing the Hidden Nature of Space and Time: Charting the Course for Elementary Particle Physics*. Board on Physics and Astronomy, The National Academies of Press, 2006.

Complex Plasma Physics

[1] H. Thomas, et al. *Physical Review Letters* 73 (1994): 652.
[2] Vladimirov, S.V., K. Ostrikov, A.A. Samarian. *Physics and Application of Complex Plasmas*. World Scientific Press, 2005.
[3] Fortov, V.E., et al. Physics Reports. 421 (2005): 1.
[4] Morfill, G.E. and A.V. Ivlev. Reviews of Modern Physics 81 (2009): 1353.

Vibrations and Granular Matter Physics

[1] Hinrichsen, Haye and Dietrich E. Wolf, eds. *The Physics of Granular Media*. Wiley-VCH Verlag GmbH & Co, 2004.
[2] Hofmeister, P., J. Blum and D. Heißelmann. "The flow of granular matter under reduced-gravity conditions." In Powders & Grains, 2009.
[3] Beysens, D. and P. Evesque. "Vibrational phenomena in near-critical fluids and granular matter." In *Topical Teamsinthe Life & Physical Sciences, Towards New Research Applications in Space*, SP 1281. Noordwijk, The Netherlands: ESA, Publication Division.
[4] Beysens, D., D. Chatain, P. Evesque and Y. Garrabos. "High-Frequency Driven Capillary Flows Speed Up the Gas–Liquid Phase Transition in Zero-Gravity Conditions". *Physical Review Letters* 95 (2005): 034502.

Dust Physics

[1] Güttler, C., M. Krause, R.J. Geretshauser, R. Speith and J. Blum. "The Physics of Protoplanetesimal Dust Agglomerates. IV. Towards a Dynamical Collision Model". The *Astrophysical Journal*. 701 (2009): 130–141.
[2] Salter, D.M., D. Heißelmann, G. Chaparro, G. van der Wolk, P. Reißaus, E. de Kuyper, P. Tuijn, R.W. Dawson, M. Hutcheon, G. Drinkwater, B. Stoll, K. Gebauer, F.J. Molster, H. Linnartz, G. Borst, H.J. Fraser and J. Blum. "A Zero-Gravity Instrument to Study Low Velocity Collisions of Fragile Particles at Low Temperatures". The Review of Scientific Instruments 80 (2009): 074501.

19

Fluid Physics

Daniel A. Beysens

CEA-Grenoble and ESPCI-Paris-Tech, Paris, France

19.1 Introduction

The investigations in fluid physics aim at predicting the new behavior of fluids in space and explaining intriguing observations. Fluids (gas, liquid) are present everywhere, that is why many problems of fluid physics are also discussed in the other fields of research discussed in this Book. Fluid behavior is actually markedly different in the space environment than on Earth. Instead of being submitted to the steady Earth gravitational acceleration, fluids in space have to face low gravity and time-dependent acceleration. Forces, as capillary forces, which are usually small or negligible and are generally ignored on Earth, become dominant and lead to unexpected and counterintuitive behavior. Others, as buoyancy forces, disappear or are greatly reduced, making other processes (diffusion, thermocapillary motion) prevailing on buoyancy-induced convection.

In the following, we give an overview of the main fields of research. More information can be found in the references and books [1–3].

19.2 Supercritical Fluids and Critical Point Phenomena

The critical point is the starting point of a new state for gas and liquids. Here pressure and temperature become high enough such that liquid and gas mix together as a dense gas, a "supercritical fluid". In the vicinity of the critical point, all fluids behave in a similar manner. Studying one fluid enables the properties of all fluids to be deduced, this is the so-called critical point universality. Kenneth G. Wilson got the Nobel Prize in 1982 for this discovery. The study of pure fluids near their critical point and, to a lower extent,

Table 19.1 Main parameters in fluid physics

Number	Definition	Meaning	Others
Capillary	$Ca = \frac{\sigma}{\eta V}$	capillary velocity/ fluid velocity	η: dynamic (shear) viscosity; V: fluid velocity
Peclet	$Pe = \frac{VL}{D}$	fluid velocity/ diffusion	D, diffusion coefficient (thermal or solutal)
Marangoni	$Ma = \frac{\sigma_T \Delta T L}{\eta D}$	surface tension forces/ viscous forces	σ_T, surface tension thermal derivative; ΔT, temperature difference; D, thermal diffusivity; L, characteristic length
Weber	$We = \frac{\rho L V^2}{\sigma}$	inertial energy/ capillary energy	ρ, fluid density

liquid mixtures near their dissolution critical point exhibits several exceptional features: very large compressibility, making fluids become compressed under their own weight on Earth, very large thermal expansion, thus even minute temperature gradients produce large density gradients. The fluid becomes subjected to strong turbulent convective flows even under extremely small temperature difference. Vapor–liquid phase transition is mostly governed by buoyancy, bubbles (or droplets) being convected upward (or downward). When nearing the critical point, the Bond number tends to zero; the closest to the critical point, the largest the influence of gravity.

The large sensitivity of near-critical fluids emphasizes the effects. Then, the utilization of the weightless environment allows data to be obtained very near their critical point. This close vicinity enables several investigations to be made [3–5].

19.2.1 Testing Universality

Pure fluids, binary liquids or polymer blends all belong to the same universality class as the Ising system for the ferromagnetism transition. A number of important properties as susceptibility and specific heat obey, asymptotically close to the critical point, universal scaling laws with universal exponents.

19.2.2 Dynamics of Phase Transition

Domains of one phase nucleate and grow at the expense of the other phase. The process, once buoyancy has been suppressed, has been shown to obey universal master laws.

19.2.3 New Process of Thermalization

The "Piston effect" has been discovered. In a closed sample submitted to a temperature rise or heat flux at a border wall, temperature rises very rapidly due to the adiabatic heating by the expansion (heating) or contraction (cooling) of the thermal boundary layer [3]. In contrast to the critical slowing down of the thermal diffusion, this process is all the more rapid as the critical point is neared, leading to "critical speeding-up" instead. On Earth, this phenomenon competes with buoyancy flows to fasten thermal equilibration.

19.2.4 Supercritical Properties

Fluids like oxygen and hydrogen are also used in space under supercritical conditions because they show up as a homogeneous fluid whatever the spacecraft or satellite accelerations are, including the absence of accelerations, and in whatever way the gravity vector is oriented. In addition, supercritical fluids show very interesting environmental properties, for instance, supercritical carbon dioxide is a very powerful (and harmless for health and environment) solvent of organic matter. It is also possible to burn dangerous wastes, like ammunitions, in supercritical water in a very efficient and safe way. Experiments in space have just started in this area [6].

19.3 Heat Transfer, Boiling and Two-Phase Flow

Heat transfer classically uses convection and phase change (condensation, evaporation). In the latter case, two-phase flow (vapor and liquid) occurs. In space, the absence of buoyancy can considerably lower the performances of heat transfer. The investigations into low gravity are then concerned with the mechanisms involved in the process: evaporation, condensation, boiling and two-phase fluid flow.

19.3.1 Two-Phase Flows

Configurations close to an industrial process are mainly considered. A two-phase loop experiment with capillary pumping is used [7]. Capillary forces efficiently pump the liquid in a porous medium, preventing the use of a mechanical device. The basic mechanisms (convection in an evaporating phase, drop evaporation, evaporation in a porous medium) are addressed, together with more technical investigations.

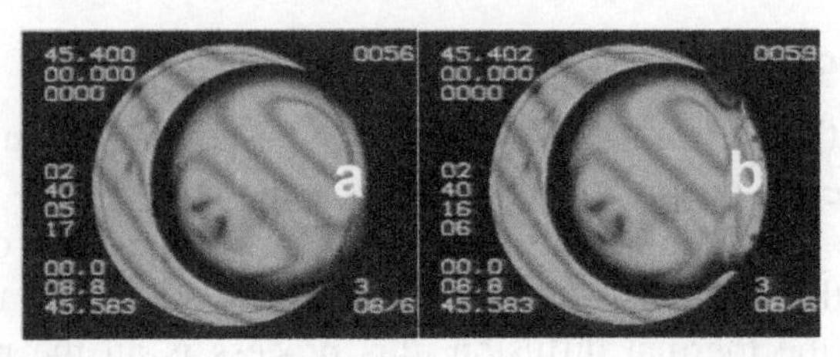

Figure 19.1 Boiling and bubble spreading under zero gravity (SF6, MIR, 1999). (a): $t = 0$, no heat flux at the wall; (b): $t = 11$ s under heat flux, vapor spreads at the contact line location due to the recoil force (from Ref. [9]).

19.3.2 Boiling and Boiling Crisis

Boiling is a highly efficient way to transfer heat. Although boiling has been studied extensively, a basic theory of boiling is still lacking, in particular concerning the phenomena very close to the heating surface. The efficiency of industrial heat exchangers increases with the heat flux. However, there is a limit called critical heat flux. It corresponds to a transition from nucleate boiling (boiling in its usual sense) to film boiling where the heater is covered by a quasi-continuous vapor film and the heat transfer efficiency drops sharply. This transition is called "burnout," "departure from nucleate boiling" or "boiling crisis".

Starting from low-gravity investigations (Figure 19.1), a vapor recoil mechanism for the boiling crisis has been proposed where a fluid molecule leaving the liquid interface causes a recoil force analogous to that created by the gas emitted by a rocket engine. At high enough heat flux, a growing bubble can forcefully push the liquid entirely away from the heating element. The evaporation is particularly strong in the vicinity of the contact line of a bubble, inside the superheated layer of the liquid [8].

19.4 Interfaces

Interfaces play a key role in many areas of science and technology: evaporation, boiling, solidification, crystallization, combustion, foam and thin film drainage, thermocapillary motion, rheology of suspensions and emulsion stability. All these domains benefit from the low-gravity environment.

19.4.1 Liquid Bridges

A liquid bridge is a volume of liquid that is surrounded by another fluid and is attached to more than one solid wall. The most studied bridge [10] is the cylindrical bridge spanning between two coaxial solid disks. Many works have been devoted to statics (equilibrium shape and stability) and dynamics. In this aspect, one must consider the behavior without thermal effects (disk vibration, rotation, stretching), with thermal gradients (thermocapillary Marangoni flows), diffusion of species (solutal Marangoni flows), phase change (unidirectional solidification—floating zone process), electric and magnetic effects (shape stabilization, convection suppression) and reactive processes (cylindrical flames).

19.4.2 Marangoni Thermo-Solutal-Capillary Flows

Capillary (Marangoni) flows develop when surface tension, σ, varies along the liquid–gas interface, from low surface tension region to high surface tension region [11]. The gradient of surface tension induces a surface flow that tends to drag the underlying bulk liquid with it. The gradient can be induced by a temperature difference, inducing a thermocapillary motion proportional to $d\sigma/dT$, or a difference in concentration, c, of a surface-active species. In this case, the motion is proportional to $d\sigma/dc$. The resulting flow velocity can be large, increasing or diminishing the buoyancy-induced flows on Earth. When liquid droplets or gas bubbles are submitted to a temperature or concentration gradient, the surface flow makes the drop or bubble move, often rapidly.

Such Marangoni flows appear naturally during phase transition where bubbles or drops migrate toward the hottest wall (if $d\sigma/dT > 0$) and convective flows appear on a solid–liquid interface. This is why most of the studies have been performed in configurations close to encountered in material sciences, for example, in the molten zone crystallization process (liquid bridge). There is indeed a direct link between the quality of crystals and the flow properties in the liquid phase. Other investigations are classically concerned with an open tank where the free surface is submitted to a temperature gradient between two parallel rigid walls.

19.4.3 Interfacial Transport

The dynamics aspect of adsorption of soluble surface-active species (surfactants) benefits from the low-gravity environment [12]. Real-time measurements of liquid–liquid and gas–liquid surface tension give a better understanding of the very nature of the interfacial transport process, in particular diffusion in the bulk toward the interface, exchange of matter and dilatational rheology of the interface.

19.4.4 Foams

A liquid foam exhibits a cellular structure made up of gas bubbles surrounded by liquid. The latter is in the form of a thin film wherever two bubbles press tightly together. The foam is not entirely static, unless it has been solidified, for example, polystyrene or metallic foams. As long as the liquid component is present, the foam evolves under the action of three processes. (i) Drainage is the motion of liquid through the foam by gravity. An equilibrium with height is reached, with "dry" foam above and "wet" foam (more than 15% liquid fraction) near the underlying liquid. (ii) Coarsening is the increase with time of the bubble size. This process is due to the diffusion of gas through the thin film as induced by the pressure difference between gas and liquid. Smaller bubbles (larger pressure) are thus eliminated in favor of the larger bubbles (lower pressure). (iii) Rupture. Coarsening ends by the rupture of thin films that causes foam to collapse.

Low-gravity experiments have led to understand the wet foam properties [13]. Some of these experiments, in addition, addressed the quite difficult technology process of metallic foam fabrication without the additives that weaken the materials elaborated on Earth.

19.4.5 Emulsions

Once vigorously shaken, immiscible liquids, like oil and vinegar or oil and water, form a dispersion of small droplets of one liquid in the other phase, that is, an emulsion. The control of the stability of emulsions is one of the most important problems in emulsion science and technology. The main factors are the following: (i) Aggregation. Different droplets of the dispersed phase aggregate in clusters. (ii) Coalescence. Two or more droplets at contact fuse together. (iii) Oswald ripening. The liquid in a small droplet diffuses to a neighboring larger droplet due to the pressure difference corresponding to the different radii of curvature. Although Brownian motion is effective to put very small drops into contact, it is mainly the gravitational forces that eventually "cream" to the surface or "settle" to the bottom.

Experiments in microgravity permit to remove the influence of gravitational forces and highlight the other destabilization causes as the dynamics of surfactant adsorption at the interface, the study of drop–drop interactions and the dynamics of phase inversion in model emulsions (from oil in water to water in oil). Experiments with metallic emulsions have shown that other processes than those described just above, like Marangoni effects, can destabilize emulsions [14].

19.4.6 Giant Fluctuations of Dissolving Interfaces

Large spatial fluctuations in concentration can take place during a free diffusion process, as the one occurring at the interface of liquids undergoing a mixing process [15]. Such fluctuations of concentration (and density) are due to a coupling between velocity and concentration fluctuations in the non-equilibrium state. As the amplitude of the fluctuations is limited by gravity, experiments indeed observe the fluctuations increase ("giant" fluctuations) when gravity is not present. The experiment is performed at the interface between two partially miscible liquids. The observation of such giant fluctuations influences other types of microgravity research, such as the growth of crystals.

19.5 Measurements of Diffusion Properties

In the concept developed by Fick law, diffusion is the tendency of species to spread uniformly in solutions. For a binary liquid solution, a component is always diffusing from high to low concentration points. It is a slow process, for example, it lasts 10^4–10^5 s to homogenize a solution over 1 cm length. When a temperature gradient is present, thermodiffusion—the "Soret effect"—takes place, with typical times on the order of diffusion. These long characteristic times mean that slow convection, as those encountered in Earth-bound measurements, can seriously perturbate the concentration field [16]. High-quality low-gravity measurements eliminate such disturbing motions inside the non-homogeneous samples. Many experiments have been conducted in materials of scientific or industrial interest as molten salts, metallic alloys and organic mixtures. They led to values that are significantly lower than those measured on Earth, leading to discriminate between the many different complex theories that aim at predicting the diffusion behavior. Reliable data are also obtained for the computer industry (solidification, crystallization of materials) and oil companies (diffusion coefficient of crude oils).

19.6 Vibrational and Transient Effects

Most experiments which are performed under space microgravity conditions are selected because of their sensitivity to gravity effects. They are thus also sensitive to acceleration variations that correspond to maneuvers, leading sometimes to unwanted sloshing motions and also to erratic or non-erratic vibrations ("g-jitters").

19.6.1 Transient and Sloshing Motions

During the cutoff or re-ignition phase of a spacecraft engine, or during the orbiting maneuvers of a satellite, the motion of two-phase fluids (e.g., oxygen and hydrogen in equilibrium with their vapor) in the reservoirs can exhibit severe sloshing motion that can even lead to stop the alimentation of the engines. The physics of the problem is complicated by the fact that the liquid-free surface is not simply flat. In addition, when the fluids and the solid mass are comparable in magnitude, the dynamics of the system are coupled [17]. Experiments to validate three-dimensional computational fluid dynamics simulation have been carried out mostly on model fluids, by using scaling with the Bond and Weber numbers, the main numbers involved in the problem. Only very few experiments were performed in real spacecrafts' tanks or with real cryogenic fluids (oxygen, hydrogen).

19.6.2 Vibrational Effects

Knowledge, prediction and minimization of vibration effects are often a necessity when dealing with the control of space experiments. Depending on the vibration amplitude and frequency, the density inhomogeneities (due to thermal gradients or vapor and liquid inclusions) tend to orientate parallel or perpendicular to the vibration direction. Vibrations applied to mechanical systems can induce destabilization or stabilization, depending on the characteristic features of the vibrations (frequency, amplitude) and the direction of vibration with respect to the density gradient orientation. Many equilibrium and non-equilibrium phenomena can then be affected by the presence of high-frequency vibrations.

Vibrations can easily provoke average motions in fluids with density inhomogeneities, counterbalancing or emphasizing the gravitational flows on Earth and inducing in space effects that are similar to those provoked by gravity. In particular, thermal instabilities similar to the one encountered on Earth, as the well-known Rayleigh–Bénard instabilities, can be induced by vibrations [18]. Vibrations can then be of interest for the management of fluids and can be considered as a mean to create an "artificial" gravity in space (see Chapter 15).

A number of investigations have been conducted under weightlessness during phase transition, especially near a vapor–liquid critical point [19]. Thermo-vibrational aspects have been the object of experiments with liquid and liquid solutions [20] to evaluate the effect of vibration on the Soret coefficients (thermodiffusion) and in homogeneous, supercritical fluids near their critical point where the effects are magnified. Experiments have also been conducted in material science (solidification). They all conclude that vibrations can induce large effects on fluid behavior.

19.7 Biofluids: Microfluidics of Biological Materials

Although the basic constituents of a biological fluid, for example blood, are of the order of micrometers, gravity and its absence can deeply affect the behavior of biofluids [21]. The other length scales (diameter and length of the vessel) are indeed large. Investigations into weightlessness of vesicles, which are good models for blood cells, show that vesicles can undergo temporal oscillations under the influence of shear flow. These studies have also industry relevance as micron-sized particles that are very close to blood compounds (red blood cells, white blood cells) can be separated by a hydrodynamic focusing method (the so-called split technique). Microgravity environment is mandatory to improve the process.

References

[1] Monti, R. *Physics of Fluids in Microgravity.* London and New York: Taylor and Francis, 2001.

[2] Seibert, G. *A World Without Gravity–Research in Space for Health and Industrial Processes.* Noordwijk: ESA Publication Division, 2001.

[3] Zappoli, B., D. Beysens, and Y. Garrabos. "Heat Transfers and Related Effects in Supercritical Fluids." Dordrecht Heidelberg New-York London: Springer, 2015.

[4] Barmatz M., and I. Hahn. "Critical Phenomena in Microgravity: Past, Present, and Future." *Review of Modern Physics* 79 (2007): 1–52.

[5] Shen, B., and P. Zhang. "An Overview of Heat Transfer Near the Liquid-Gas Critical Point Under the Influence of the Piston Effect: Phenomena and Theory." *International Journal of Thermal Sciences* 71 (2013): 1–19.

[6] Pont, G., S. Barde, B. Zappoli, Y. Garrabos, C. Lecoutre, D. Beysens, M. Hicks, U. Hegde, I. Hahn, N. Bergeon, B. Billia, R. Trivedi, and A. Karma. "DECLIC, now and tomorrow." *Proceedings of 64th International Astronautical Congress (IAC)*, Pékin, Chine, (23–27 Sept. 2013), IAC-13,A2,5,5,x19013.

[7] Lebaigue, O., C. Colin, and A. Larue de Tournemine. "Forced Convection Boiling and Condensation of Ammonia in Microgravity." *Annals of [20]*

the Assembly for International Heat Transfer Conference 13 (2006): 250–259.

[8] Nikolayev, V.S., D. Chatain, Y. Garrabos, and D. Beysens. "Experimental Evidence of the Vapour Recoil Mechanism in the Boiling Crisis." *Physical Review Letters* 97 (2006): 184503-1–184503-4.

[9] Garrabos, Y., C. Chabot, R. Wunenburger, J.-P. Delville, and D. Beysens. "Critical Boiling Phenomena Observed in Microgravity." *Journal de Chimie Physique* 96 (1999): 1066–1073.

[10] Martinez, I., and J. Perales. "Mechanical behavior of liquid bridges in microgravity." In *Physics of Fluids in Microgravity*, edited by R. Monti, 21–45. London and New York: Taylor and Francis, 2001.

[11] Castagnolo, D., and R. Monti. "Thermal Marangoni flows." In: *Physics of Fluids in Microgravity*, edited by R. Monti, 78–125, London and New York, NY: Taylor and Francis, 2001.

[12] Passerone, A., L. Liggieri, and F. Ravera. "Interfacial Phenomena." In *Physics of Fluids in Microgravity*, edited by R. Monti, 46–77. London and New York: Taylor and Francis, 2001.

[13] Saint-Jalmes, A., S. Marze, H. Ritacco, D. Langevin, S. Bail, J. Dubail, G. Roux, L. Guingot, L. Tosini, and P. Sung. "Diffusive Liquid Transport in Porous and Elastic Materials: The Case of Foams in Microgravity." *Physical Review Letters* 98 (2007): 058303-1–058303-4.

[14] Miller, R., and L. Liggieri. *Interfacial Rheology*. Leiden: Brill, 2009.

[15] Vailati, A., and M. Giglio. "Giant fluctuations in a free diffusion process." *Nature* 390 (1997): 262–265.

[16] Van Vaerenbergh, S., and J.-C. Legros. In *Physics of Fluids in Microgravity*, edited by R. Monti, 178–216. London and New York: Taylor and Francis, 2001.

[17] Vreeburg, J.P.B., and A.E.P. Veldman. "Transient and sloshing motions in an unsupported container." In *Physics of Fluids in Microgravity*, edited by R. Monti, 293–321. London and New York, NY: Taylor and Francis, 2001.

[18] Gershuni, G.Z., D.V. Lyubimov, "Thermal Vibrational Convection." New York, NY: Wiley, 1998.

[19] Beysens, D., D. Chatain, P. Evesque, and Y. Garrabos. "High Frequency Driven Capillary Flows Speed Up the Gas-Liquid Phase Transition in Zero-Gravity Conditions." *Physical Review letters* 95 (2005): 034502-1–034502-4.

[20] Mialdun, A., I.I. Ryzhkov, D.E. Melnikov, and V. Shevtsova. “Experimental Evidence of Thermovibrational Convection in Low Gravity.” *Physical Review Letter* 101 (2008) 084501-1–084501-4.
[21] Mader, M., C. Misbah, and T. Podgorski. “Dynamics and Rheology of Vesicles in a Shear Flow Under Gravity and Micro-Gravity.” *Microgravity Science and Technology* 18 (2006): 199–2003.

20

Combustion

Christian Chauveau

CNRS–INSIS–ICARE, Orléans, France

20.1 Introduction

Combustion is a rapid, self-sustaining chemical reaction that releases a significant amount of heat and as such involves elements of chemical kinetics, transport processes, thermodynamics and fluid mechanics.

Combustion is a key element in many technological applications in modern society. In itself, combustion is one of the most important processes in the world economy. Combustion underlies almost all systems of energy generation, domestic heating and transportation propulsion. It also plays a major role at all stages in the industrial transformation of matter, ranging from the production of raw materials to the complex assembly of industrial products. Although combustion is essential to our current way of life, it poses great challenges to society's ability to maintain a healthy environment and to preserve vital resources for future generations. Improved understanding of combustion will help us to deal more effectively with the problems of pollution, global warming, fires, accidental explosions and the incineration of dangerous waste. In spite of extensive scientific research since more than a century, many fundamental aspects of combustion are still poorly understood.

The objectives of scientific research on microgravity combustion are initially to increase our knowledge of the fundamental combustion phenomena that are affected by gravity, then to use the research results to advance science and technology related to combustion in terrestrial applications and finally to tackle questions of security related to fires on board spacecraft. The following review articles [1–5] are to be consulted, for those who would be interested to look further into the role of microgravity in combustion research.

20.2 Why Combustion Is Affected by Gravity?

Microgravity combustion scientists undertake experiments both in ground-based microgravity facilities and in orbiting laboratories and study how flames behave under microgravity conditions.

Microgravity research allows the conduct of new experiments in which buoyancy-induced flows and sedimentation are virtually eliminated. Combustion usually involves large temperature increases resulting in a consequent reduction in density, ranging from a factor of two to ten depending on the situation. As a result of this density change, the combustion processes in normal gravity are usually strongly influenced by natural convection. The rise of hot gas creates a buoyancy-induced flow favoring gas mixing from the fuel, oxidizer and combustion products. Under conditions of reduced gravity, natural convection is cancelled (or greatly reduced), and therefore the characteristics of combustion processes can be profoundly altered.

The reduction of buoyancy-induced flows has several features that are particularly useful for fundamental and applied scientific research on combustion. By eliminating the effects of natural convection, a quiescent environment is created, conducive to more symmetrical results. This facilitates comparisons with numerical modeling results and with theories. Furthermore, the elimination or drastic reduction of buoyancy-induced flows can reveal and highlight weaker forces and flows that are normally masked, such as electrostatics, thermocapillarity and diffusion. Lastly, the elimination of disturbances caused by buoyancy forces can increase the duration of experiments, thus allowing the examination of the phenomena over larger time scales.

For purposes of simplification, the numerical models developed in combustion research often assumed that the mixture of the initial components is homogeneous. Sedimentation affects combustion experiments involving drops or particles, since the components with the highest density will be driven down into the gas or liquid, and hence their movement relative to other particles creates an asymmetric flow around the falling particle. The presence of these concentration gradients in the mixture before combustion complicates the interpretation of experimental results. In normal gravity conditions, experimenters must implement devices to stabilize and homogenize dispersed media, for example, supports, levitators or stirring devices. In microgravity, gravitational settling is nearly eliminated, allowing the stabilization of free droplets, particles, bubbles, fog and droplet networks for fundamental studies on ignition and combustion in heterogeneous media.

To date, scientific research in combustion has shown major differences in the structure of different flames burning either in microgravity or under normal gravity. Besides the practical implications of these results in terms of combustion efficiency, pollutant control and flammability, these studies have established that a better understanding of the sub-mechanisms involved in the overall combustion process is possible by comparing the results obtained in microgravity with those obtained in normal gravity.

20.3 Reduced Gravity Environment for Combustion Studies

While microgravity is the operational environment related to Earth-orbiting space laboratories, it is important to note that "ground-based facilities" allowing gravity reduction also serve the scientific community and enable relevant combustion studies to be carried out. Experiments conducted in suborbital sounding rockets, during the parabolic trajectory of an aircraft laboratory, and free-fall drop towers, significantly complement the limited testing opportunities available aboard the International Space Station. In fact, the contributions of research conducted in these so-called ground-based microgravity facilities have been essential to the acknowledged success of microgravity combustion research. These helpful facilities allow us to consider microgravity as a tool for combustion research, in the same way as an experimenter can vary pressure or temperature; microgravity can also act on the gravitational acceleration parameter. Additional contributions of high value to microgravity combustion research come from "normal-gravity" reference ground tests and from analytical modeling.

Some results can be cited here to illustrate how so many combustion processes are affected by gravity. As an example of spectacular results, stationary premixed spherical flames (i.e., flame balls), whose existence was predicted by theory but had never been confirmed by any experiments in normal gravity, were observed uniquely in microgravity [6]. Recently, flame extinguishment experiment conducted on droplet combustion has demonstrated radiative and diffusive extinction, combustion instabilities, lower flammability limits and unexplained vaporization after visible flame extinction. This behavior ever brought back before leads to the possible existence of cool-flame chemistry [7]. The inhibition of flame spreading along both solid and liquid surfaces is of primary importance in fire safety. Experimental studies have revealed major differences between normal and reduced gravity conditions, concerning the ignition and flame spreading characteristics of solid and liquid fuels.

While most microgravity combustion experiments have been conducted in dedicated and unique experimental apparatus, the recent commissioning of the Combustion Integrated Rack (CIR) in the FCF experimental rack (Fluid Combustion Facility) aboard the ISS should enable more investigators to have access to this microgravity environment [8].

20.4 Conclusions

Compared to experimental combustion studies in laboratories, the number of microgravity experiments is small. Nevertheless, important discoveries have already emerged from microgravity combustion investigations. The numerous facilities existing now, both "ground-based" microgravity facilities and on board the ISS, made available to the scientific community by space agencies, suggest that new microgravity combustion experiments will significantly advance fundamental understanding in combustion science. It is hoped that this will help to maintain a healthy environment and preserve vital resources for future generations.

References

[1] Law, C.K., and G.M. Faeth. "Opportunities and Challenges of Combustion in Microgravity." *Progress in Energy and Combustion Science* 20, no. 1 (1994): 65–113.

[2] Williams, F.A., "Combustion Processes Under Microgravity Conditions." In *Materials and Fluids Under Low Gravity, Lectures Notes in Physics*, edited by L. Ratke, H. Walter, and B. Feuerbacher, pp. 387–400, Berlin: Springer, 1995.

[3] Microgravity Combustion Science: 1995 Program Update, Technical Memorandum NASA/TM-106858, 1995.

[4] Ronney, P.D. "Understanding Combustion Processes Through Microgravity Research." *Symposium (International) on Combustion* 27, no. 2 (1998): 2485–2506.

[5] Friedman, R., S.A. Gokoglu, and D.L. Urban. Microgravity combustion research: 1999 Program and results, Vol. NASA/TM-1999-209198, NASA, Glenn Research Center, Cleveland, Ohio, 1999.

[6] Ronney, P.D., M.S. Wu, H.G. Pearlman, and K.J. Weiland. "Experimental Study of Flame Balls in Space: Preliminary Results from STS-83." *Aiaa Journal* 36, no. 8 (1998): 1361–1368.

[7] Nayagam, V., D.L. Dietrich, P.V. Ferkul, M.C. Hicks, and F.A. Williams. "Can Cool Flames Support Quasi-Steady Alkane Droplet Burning?" *Combustion and Flame* 159, no. 12 (2012): 3583–3588.

[8] O'Malley, T. F., and K.J. Weiland. The FCF Combustion Integrated Rack: Microgravity Combustion Science Onboard the International Space Station, WU-398-20-0C-00, Vol. NASA/TM—2002-210981, Cape Canaveral, FL; United States 2001.

21

Materials Science

Hans-Jörg Fecht

Ulm University, Ulm, Germany

21.1 Introduction

Materials science is an interdisciplinary field dealing with the properties of matter and its applications to various areas of science and engineering. This science investigates basically the relationship between the structure of materials and their various properties. It includes elements of applied physics and chemistry, as well as chemical, mechanical and electrical engineering. With significant attention to nanoscience and nanotechnology in recent years, materials science has been propelled to the forefront [1].

The research—from a fundamental and applied point of view—is concerned with the synthesis, atomic structure, chemical element distribution and various favorable properties of materials and structures. The basic understanding and optimization of properties and structures is further supported by sophisticated computer models from the nano- to the macroscale and leads to a manifold of applications in industrial products. Prominent examples are found in the aerospace, automotive, biomedical, energy and microelectronics industries.

The interactive feedback between experiments and sophisticated computer simulations developed within the last ten years that now drives the design and processing of materials is reaching performances never been seen in the past. Thus, it becomes possible to control and optimize the defect and grain structure at critical patches of components. In this regard, two major aspects are most essential:

- the reliable determination of the **thermophysical properties** of metallic melts for industrial process design in order to understand the fundamentals of complex melts and their influence on the nucleation and growth of ordered phases, together with

- the reliable determination of the formation and selection mechanisms at microstructure scales in order to develop **new materials, products and processes**.

As a result, fresh insights into alloy solidification/processing can be gained with the potential of producing novel materials and structures, i.e., materials processed and designed in space [2, 3].

21.2 Scientific Challenges

Casting is a non-equilibrium process by which a liquid alloy is solidified. The liquid–solid transition is driven by the departure from thermodynamic equilibrium where no change can occur. From the standpoint of physics, casting thus belongs to the vast realm of out-of-equilibrium systems, which means that rather than growing evenly in space and smoothly in time, the solid phase prefers to form a diversity of microstructures.

Actually, the relevant length scales in casting are widespread over 10 orders of magnitude. At the nanometer scale, the atomic processes determine the growth kinetics and the solid–liquid interfacial energy, and crystalline defects such as dislocations, grain boundaries and voids are generally observed. Macroscopic fluid flow driven by gravity or imposed by a stimulus (electromagnetic field, vibration, etc.) occurs in the melt at the meter scale of the cast product. The characteristic scales associated with the solidification microstructures are mesoscopic, i.e., intermediate, ranging from dendrite tip/arm scale (1–100 μm) to the grain size (mm–cm). It follows that the optimization of the grain structure of the product and inner microstructure of the grain(s) during the liquid-to-solid phase transition is paramount for the quality and reliability of castings, as well as for the tailoring of new advanced materials for specific technological applications.

On this basis, the quantitative numerical simulation of casting and solidification processes is increasingly demanded by manufacturers, compared to the well-established but time-consuming and costly trial-and-error procedure. It provides a rapid tool for the microstructural optimization of high-quality castings, in particular where process reliability and high geometric shape accuracy are important (see, for example, Figure 21.1 exhibiting cast structural components and the temperature distribution during casting of a car engine block). Any improvement of numerical simulation results in an improved control of fluid flow and cooling conditions that enables further optimization of the defect and grain structure as well as mechanical stress distribution.

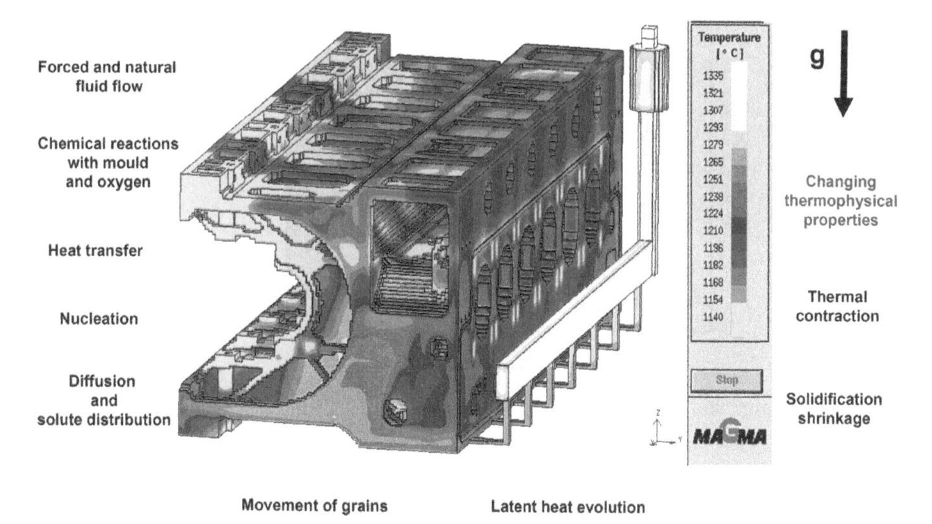

Figure 21.1 A wide range of fundamental events during casting of complex components, here a car engine with varying local temperatures.

During processing from the melt crystal nucleation and growth is in most situations the first step achieved by cooling of liquid below its thermodynamic equilibrium solidification (liquidus) temperature. Alternatively, when the formation of nuclei fails, or the growth of nuclei is very sluggish, there is formation of a metallic glass at the glass transition temperature. If the crystal nucleation rate is sufficiently low and if the growth of nuclei is sufficiently slow over the entire range/state of the undercooled liquid below the liquidus temperature, eventually the liquid freezes to a non-crystalline solid—a glass.

In particular, new metallic glasses that can be produced in large dimensions and quantities now—the so-called bulk metallic glass or super metals—are emerging as an important industrial and commercial material, superior to conventional Ti, Al- or Fe-based alloys. They are characterized by several times the mechanical strength (up to 5 GPa) in comparison with conventional materials, excellent wear properties and corrosion resistance due to the lack of grain boundaries (see Figures 21.2, 21.3).

21.3 Specifics of Low-Gravity Platforms and Facilities for Materials Science

Fresh insight into the stable and metastable undercooled liquid state and metallic alloy solidification can be gained with the potential of engineering

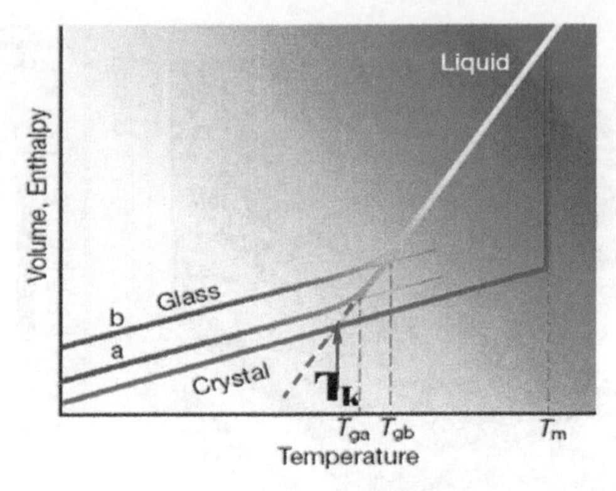

Figure 21.2 Volume and enthalpy of a glass-forming alloy as a function temperature and undercooling.

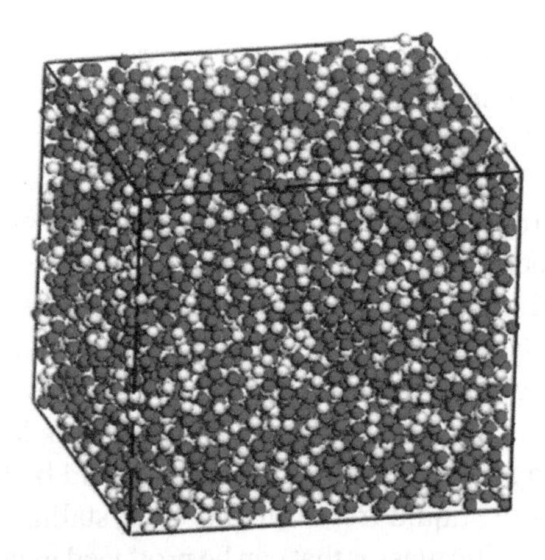

Figure 21.3 Atomic structure in an MD simulation of a Or-Cu glass [1].

novel microstructures. To perform these experiments, it is important to have access to extended periods of reduced gravity.

21.3.1 Parabolic Flights

Parabolic flights generally provide about 20 seconds of reduced gravity. For materials science experiments at high temperatures, this time is barely sufficient for melting, heating into the stable liquid and cooling to solidification of most metallic alloys of interest in a temperature range between 10,000–2,000°C. Surface oscillations can be excited by a pulse of the heating field, and the surface tension and viscosity are obtained from the oscillation

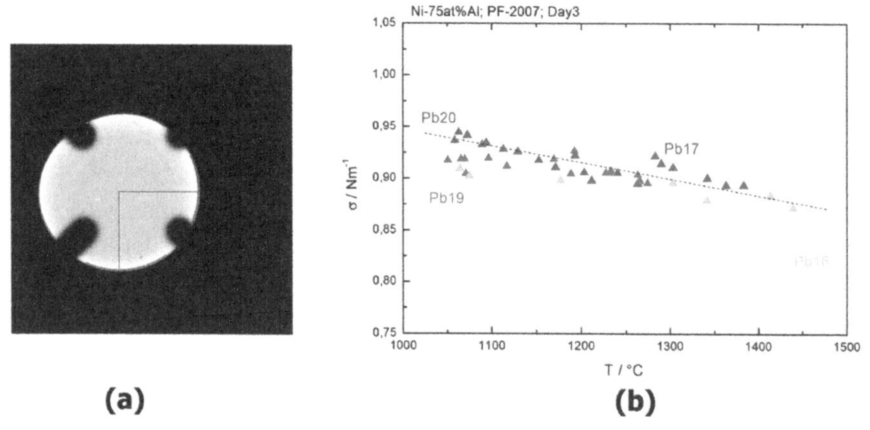

Figure 21.4 (a) Video image of a fully spherical liquid sample of a NiAl alloy in EML obtained on a parabolic flight for surface tension and viscosity measurements of liquid metallic alloys, (b) Surface tension of a drop of molten Ni-75 at.% Al.

frequency and damping time constant of the oscillations, respectively. Processing must be performed in a gas atmosphere under convective cooling conditions. Under these conditions, however, thermal equilibrium of the melt cannot be reached.

Using the Electromagnetic Levitator, surface oscillations of the liquid hot drop with diameter of 8 mm (Figure 21.4a) can be, for instance, introduced by an electromagnetic pulse and the results analyzed by a high-speed high-resolution video camera. Among the results, Figure 21.4b shows the surface tension as a function of temperature in the range 1050–1450°C of liquid Ni–75 at.% Al processed under low gravity on four parabolic flights with 10 seconds of processing time each.

21.3.2 TEXUS Sounding Rocket Processing

TEXUS sounding rockets offer a total of 320 seconds of reduced gravity which is typically split between two different experiments. As compared to parabolic flights, the microgravity quality is improved by far. Stable positioning and processing of metallic specimen with widely different electrical resistivity and density was achieved in different sounding rocket flights with an adapted electromagnetic levitation device. In Figure 21.4, the temperature-time profile of the Fe alloy processed successfully in TEXUS 46 EML-III is shown as an example. However, also under these conditions thermal equilibrium or steady-state conditions of the liquid phase cannot be attained due to the limitations in processing time.

21.3.3 Long-Duration Microgravity Experiments on ISS

A series of microgravity research is now commencing onboard the ISS in a number of multiuser facilities afforded by major space agencies, such as the European Space Agency (ESA) in the Columbus module. The short microgravity time on parabolic flights and TEXUS sounding rocket experiments necessitates forced convective cooling to solidify the specimen before the end of the microgravity time. As a consequence, long-duration microgravity measurements are needed for accurate calorimetric and thermal transport property measurements over a large temperature range. It can be expected that the following properties necessary for a full and thorough analysis of solidification processes will be performed at varying temperatures (ThermoLab-ISS program: an international effort sponsored by several national agencies including ESA, DLR, NASA, JAXA, SSO).

Furthermore, in Materials Science both directional and isothermal solidification experiments will be performed in the Materials Science Laboratory (MSL, see Figure 21.5) using dedicated furnace inserts, with the possibility of applying external stimuli such as a rotating magnetic field to force fluid low in a controlled way. The electromagnetic levitator (EML) will enable

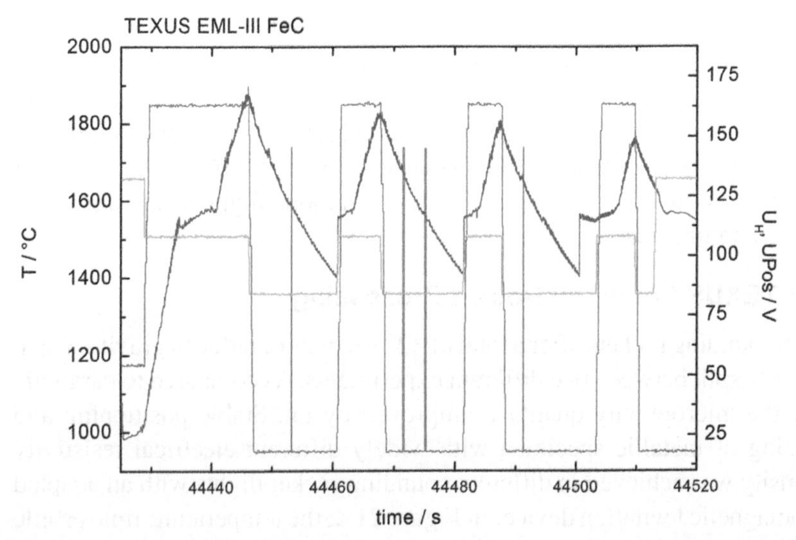

Figure 21.5 Temperature-time profile of Fe-C alloys processed on the TEXUS 46 EML-III sounding rocked flight with temperature scale (left) and heater and positioner voltage (right ordinate).

containerless melting and solidification of alloys and semiconductor samples. The EML is equipped with highly advanced diagnostic tools (Figure 21.6).

21.4 Materials Alloy Selection

For a wide range of new products in the industrial production chain, solidification processing of metallic alloys from the melt is a step of uppermost importance. On this basis, several types of alloys have been selected for the following fields of usage, for example, turbine blades for land-based power plants and for jet engines sustaining high temperatures and high stress levels, low-emission energy-effective engines for cars and aerospace, functional materials with improved performance, the so-called super metals (bulk metallic glass) with ultimate strength to weight ratio and new low-weight and high-strength materials for space exploration and future space vehicles.

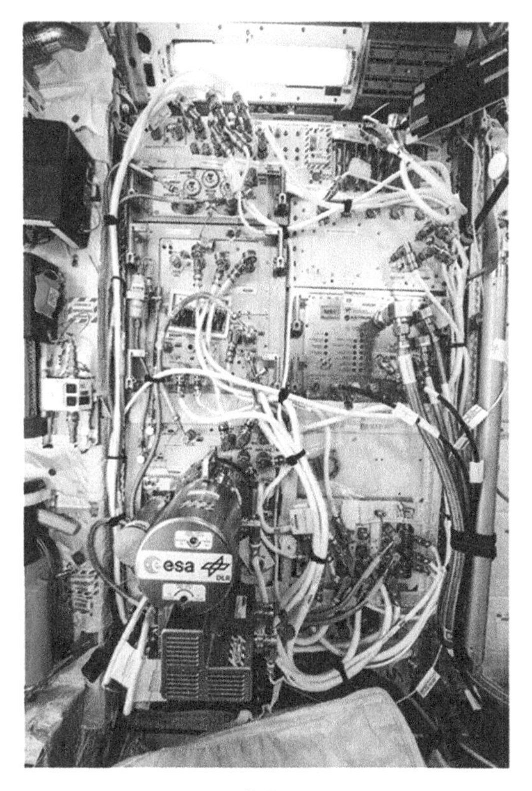

(a)

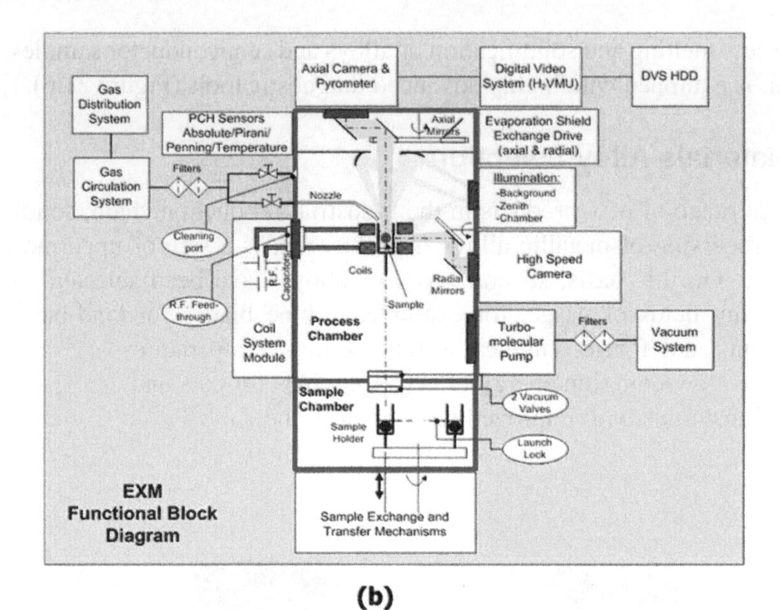

(b)

Figure 21.6 Schematic presentation of (a) the Materials Science Laboratory, and (b) the electromagnetic levitator reaching temperatures up to 2200 C.

Acknowledgements

The financial support from ESA and DLR and the scientific cooperation within the International *ThermoLab Team* (MAPs and Topical Teams) is gratefully acknowledged.

References

[1] Beysens, D., L. Carotenuto, J. vanLoon, and M. Zell, eds. *Laboratory Science with Space Data*. Berlin: Springer, 2011 (ISBN 978-3-642-21144-7).

[2] Fukuyama, H., and Y. Waseda. eds. *Advances in Materials Research, High-Temperature Measurements of Materials*, Berlin: Springer, 2009 (ISBN 978-3-540-85917-8).

[3] Fecht, H.-J., and B. Billia. “ASM Handbook.” In *Metals Process Simulation*, edited by D. Furrer, and S.L. Semiatin, Vol. 22B. Materials Park: ASM International, 2010.

6

Current Research in Life Sciences

22

Microbiology/Astrobiology

Felice Mastroleo and Natalie Leys

Belgian Nuclear Research Center (SCK•CEN), Mol, Belgium

Since the advent of space flight and the establishment of long-duration space stations in Earth's orbit, such as Skylab, Salyut, Mir, and the ISS, the upper boundary of our biosphere has extended into space. Such space missions expose humans and any other organisms to living conditions not encountered on Earth.

22.1 Radiation Environment

Life on Earth, throughout its almost four billion years of history, has been shaped by interactions of organisms with their environment and by numerous adaptive responses to environmental stressors. Among these, radiation, both of terrestrial and of cosmic origin, is a persistent stress factor that life has to cope with [7]. Radiation interacts with matter, primarily through the ionization and excitation of electrons in atoms and molecules. These matter–energy interactions have been decisively involved in the creation and maintenance of living systems on Earth. Because it is a strong mutagen, radiation is considered a powerful promoter of biological evolution on the one hand and an account of deleterious consequences to individual cells and organisms (*e.g.*, by causing inactivation or mutation induction) on the other.

In response to harmful effects of environmental radiation, life has developed a variety of defense mechanisms, including the increase in the production of stress proteins, the activation of the immune defense system, and a variety of efficient repair systems for radiation-induced DNA injury. Radiobiologists have long believed that ionizing radiation, such as gamma rays, kills cells by shattering DNA. Recently, Daly [12] and Frederikson [17] showed that proteins—not DNA—are the most sensitive targets, at least in some radiation-sensitive bacteria. In cells, oxidatively damaged DNA repair enzymes generated by sublethal ionizing radiation doses would be expected to passively promote mutations by misrepair. Oxidized proteins, however, might also actively promote mutation by transmitting damage to other cellular constituents, including DNA [16, 36].

22.2 Change in Gravity Environment

Microbes have the ability to sense and respond to mechanical stimuli. The response of microbes to certain mechanical stimuli has profound effects on their physiology [19, 38, 39, 46, 52]. The response of a cell to mechanical stimulation, such as stretch or shear force, is called mechano-adaptation and is important for cell protection in both prokaryotes and eukaryotes [19, 24, 25]. A great deal of progress has been made in understanding certain aspects of microbial mechano-adaptation, for example, mechanisms used by bacteria to respond to changes in osmotic gradients [9, 19, 41]. Studies have also documented that microbes can sense and respond to changes in culture conditions when grown in the buoyant, low-fluid-shear environment of microgravity [14, 26, 38, 39, 52]. It has been hypothesized that cells sense changes in mechanical forces, including shear and gravity, at their cell surface [24, 46].

Mechanical culture conditions in the quiescent microgravity environment of space flight are characterized by significant reductions in fluid shear [20]. This is because convection currents are essentially absent in microgravity [27].

The most commonly used microgravity simulator is the rotating wall vessel (RWV) culture apparatus (Synthecon; Texas, USA) developed by the NASA biotechnology group at Johnson Space Centre in Texas. This apparatus consists of a rotor, a culture vessel, and a platform on which the vessel is rotated. The RWV has separable front and back faces; the front face contains two sampling ports, and the back is provided with a semipermeable membrane for aeration. The assembled vessel is filled to capacity (zero headspace) with medium and inoculum, and air bubbles are removed to eliminate turbulence and ensure a sustained low-shear environment (<0.01 Pa).

In the vessel rotating around a horizontal axis, the liquid moves as a single body of fluid in which the gravitational vector is offset by hydrodynamic, centrifugal, and Coriolis (circular movement) forces resulting in the maintenance of cells in a continuous suspended orbit. In fact, this system "confuses" the biosystems (e.g., cells growing culture) perception of gravity's direction. By placing cells along the axis of rotation and spinning them perpendicular to the gravity vector, they rotate through the vector. Because the cell spins at a constant rate and gravity remains constant, the gravity vector is nulled from the cell's perspective [21]. Thus, the RWV does not generate microgravity as on the ISS, rather it randomizes gravity vectors and mimics the low turbulence of a space environment. Since the RWV apparatus provides a low-shear culture environment that simulates the aspects of space (and therefore "models microgravity"), Nickerson et al. [39] have adopted the terminology low-shear modeled microgravity (LSMMG) to refer to the RWV culture environment.

Albrecht-Buehler [1] suggested that reduced gravity suppresses buoyancy-driven convection and thus limiting the mechanism of mixing of fluid to diffusion. Along similar lines, McPherson [34] suggested that the lack of convective mixing under reduced gravity conditions created a quiescent environment that resulted in a "depletion zone" around a growing protein crystal, which favored the formation of a crystal with better quality. Based on these studies, it was hypothesized that this same type of phenomenon might occur around a growing bacterial cell under reduced gravity conditions [28, 45]. Similarly, a few studies have speculated that bacteria may indirectly respond to reduced gravity conditions because of changes in their immediate environment resulting either from changes in mass diffusion or from other chemical alterations, such as accumulation of toxic by-products [28], or limitations in the availability of nutrients [3–5, 52]. Gene expression studies performed on *Escherichia coli* K12 under modeled reduced gravity conditions support the hypothesis that bacteria are actively responding to the changes in nutrient availability, imposed by the altered mass transport under these conditions [49, 52]. Creation of zones of nutrient depletion over-time in their immediate surroundings makes these bacteria respond in a way that is similar to their entrance into stationary phase (starvation). Stationary phase cells are generally characterized by the expression of starvation inducible genes and genes associated with multiple stress responses. Microgravity-exposed bacteria appeared, for example, better able to handle subsequent stressors including osmolarity, pH, temperature, and antimicrobial challenge. More recently it was reported that the fluid quiescence and reduced mixing could enhance the accumulation of quorum sensing (QS) molecules in the bacterium's surroundings and thus promoting QS-related gene expression, independently of change in cell concentration [22, 31].

Another microgravity simulator the random positioning machine (RPM) or three-dimensional clinostat is a laboratory instrument to randomly change the position of an accommodated (biological) experiment in three-dimensional space [23]. The layout of the RPM consists of two frames and experiment platform. The frames are driven by means of belts and two electromotors. Both motors are controlled on the basis of feedback signals generated by encoders, mounted on the motor-axes, and by "null position" sensors on the frames. On the RPM, the samples are fixed as close as possible to the center of the inner rotating frame. This frame rotates within another rotating frame. The RPM has been extensively use to study cytoskeleton structure and motility of human cells [35, 53] and plant gravitropism [6, 23] and more recently the RPM has been used to study bacterial cultivation [8, 13, 30–33]. Comparing LSMMG to RPM cultivation, the two simulators appeared to induce a similar response in *Rhodospirillum rubrum* [31] while *Pseudomonas aeruginosa* only responded to cultivation in LSMMG compared to the control conditions [11] and *Cupriavidus metallidurans* proteome was highly affected only by RPM cultivation [30]. Therefore, one should be cautious to conclude which of the two simulators induced the higher response when cultivating bacteria.

The use of magnetic levitation has also been introduced to balance the force of gravity on a levitating object [10, 50]. However, a major constrain in using diamagnetic levitation is the requirement for large magnetic fields gradients at the levitation point that may influence biological systems. In addition, oxygen dissolved in the liquid culture medium is similarly attracted by the magnetic field [2]. Since oxygen in the liquid is consumed by the bacteria and replaced at the liquid–air interface (from the oxygen in the air above the liquid), an oxygen concentration results, producing a corresponding gradient in the magnetic force density that can cause convection in the liquid medium. Therefore, for diamagnetic levitation to be a useful model of space-related microgravity, where density-driven convective transport is absent, paramagnetically driven convection of oxygen should be prevented. This could be achieved by performing experiment in anaerobic conditions or in nonliquid culture [15]. Beuls et al. [8] compared 3 microgravity simulators, LSMMG, RPM, and diamagnetic levitation, and found no differences in the capacity of *Bacillus thuringiensis* to perform plasmid transfer compared to the control condition. In this case of Gram-positive bacterium, this ability to exchange plasmids in microgravity, as efficiently as occurring on Earth, could be seen as highly relevant in the frame of potentially increasing antibiotic resistances and bacterial virulence in space [8].

22.3 Space Flight Experiments and Related Ground Simulations

In general, one could conclude that space flight has been shown not to hinder bacterial growth, on the contrary, it can enhance the growth of planktonic bacterial cultures, possibly through its influences on fluid dynamics [28].

Biofilm formation can also occur in microgravity [33, 42], an issue that, for example, must be addressed during the design of air and water recycling systems for long-term space flights [40].

Various experiments suggested that antibiotics are less efficient in space flight conditions [29, 48]. Bacteria exposed to space flight stresses may become more resistant to antibiotics over a short, introductory period, while losing most but not all of that resistance over the long term [29]. For the future, this possibly changing response of bacteria to antibiotics in space flight may imply that disinfection may be problematic. In addition, it may be difficult to treat an illness or injury with antibiotics on short-term missions, due to the tendency of bacteria to resist them. On long-term missions, such as periods spent on space stations or trips to other planets, it may be difficult to predict the response of bacteria to a certain antibiotic. While most bacteria seem to become more susceptible to antibiotics after long-term space flight exposure [29], a few may retain resistance, leading to potential hazard for the all crew.

Recently, Wilson et al. [54] provided the first direct evidence that growth during space flight can alter the virulence of a pathogen; in this study, *Salmonella enterica* serovar Typhimurium grown in space flight displayed increased virulence in a murine infection model compared with identical ground controls. Importantly, these results correlate with previous findings in which the same strain displayed increased virulence in the murine model after growth in the low-shear microgravity-like conditions of the RWV bioreactor [37, 55]. In agreement with the increased virulence observed for the space flight samples, bacteria cultured in flight exhibited cellular aggregation and extracellular matrix formation consistent with biofilm production. Moreover, several *Salmonella* genes associated with biofilm formation changed expression in flight [54].

Very few attempts were made to mimic space-ionizing radiation on ground and compare it to actual space flight experiment involving microorganisms. Rea et al. [43] studied the effect of ionizing radiation on photosynthetic organisms including the cyanobacterium *Arthrospira platensis* that appeared to maintain the highest photosynthetic efficiency in flight experiments. The authors concluded that in space, the effect of ionizing radiation is enhanced compared to that observed in ground facilities with a single beam of radiation. Our group had a more complete approach trying to match the actual dosimetry measurements inside the ISS meaning about 2 mGy over 10 days [18] at the time of our space experiment involving *R. rubrum* on agar plates and combining it with ground simulation of microgravity using the RPM [32].

We could put forward the importance of medium composition and culture setup on the response of the bacterium to space flight-related environmental conditions but low overlap was obtained for both the microgravity simulation and the ionizing radiation experiments compared to the space flight experiments.

One must be aware that space experiments are always subject to the inconveniences of access to space. Space biology researchers face many limitations; include sample preparation long before the flight with prolonged storage, a strictly limited number of samples and repetitions, strong acceleration during take-off, and a second storage period before recovery and analysis of the samples. In addition, during space flight cells are exposed to many more changing factors than just the reduced gravity (e.g., increased gravity/acceleration during launch and landing, increased radiation doses, different electromagnetic fields, pressures changes, enclosed environment) and ionizing radiation. These constraints always impose a certain degree of caution when drawing conclusions on the effects of space on cells and organisms [45]. It could be therefore difficult to detect the subtle effects caused by the low dose of space radiation inside the ISS while drastic effects on liquid samples due to change in gravity conditions could be easily put forward. As a consequence, these studies did indicate that the effects observed in space flight experiments are partially (potentially even largely) due to the low-shear environment typical of the space environment.

Contrary to open environments, confinement conditions can influence the prevalence, ecology and diversity of the microbial communities via unusual conditions of atmospheric humidity, water condensation, or accumulation of biological residues [51]. Confined habitats such as Antarctic Concordia Station are used as a model environment for long-duration space flights to study human adaptation to isolated and confined extreme environmental situations as they allow to map and monitor the dynamics of airborne bacteria over a certain period of time. In a recent study, Shiwon et al. [44] detected resistances of up to five antibiotics in several *staphylococcal* and *enterococcal* strains from ISS and Concordia. On the other hand, Timmery et al. [47] identified putative pathogens able to perform horizontal gene transfer and potentially able to acquire new DNA and sharing genetic material in Concordia. Because most of the microorganisms originate from the crew, continuous evaluation of the bacterial ecological status in such confined environment was highly recommended [47].

References

[1] Albrecht-Buehler, G. "Possible Mechanisms of Indirect Gravity Sensing by Cells." ASGSB Bull 4, no. 2 (1991): 25–34.

[2] Aoyagi, S., et al. "Control of Chemical Reaction Involving Dissolved Oxygen Using Magnetic Field Gradient." Chemical Physics 331, no. 1 (2006): 137–141.

[3] Baker, P.W., and L. Leff. "The Effect of Simulated Microgravity on Bacteria from the Mir Space Station." Microgravity Science and Technology 15, no. 1 (2004): 35–41.

[4] Baker P.W., and L. Leff. "Mir Space Station Bacteria Responses to Modeled Reduced Gravity Under Starvation Conditions." Advance in Space Research 38 (2006): 1152–1158.

[5] Baker, P.W., M.L. Meyer, and L.G. Leff. "Escherichia Coli Growth Under Modeled Reduced Gravity." Microgravity Science and Technology 15, no. 4 (2004): 39–44.

[6] Barjaktarovic, Z., et al. "Changes in the Effective Gravitational Field Strength Affect the State of Phosphorylation of Stress-Related Proteins in Callus Cultures of Arabidopsis Thaliana." Journal of Experimental Botany 60, no. 3 (2009): 779–789.

[7] Benton, E.R., and E.V. Benton. "Space Radiation Dosimetry in Low-Earth Orbit and Beyond." Nuclear Instruments and Methods in Physics 184, no. 1–2 (2001): 255–294.

[8] Beuls, E., et al. "Bacillus Thuringiensis Conjugation in Simulated Microgravity." Astrobiology 9, no. 8 (2009): 797–805.

[9] Blount, P., and P.C. Moe. "Bacterial Mechanosensitive Channels: Integrating Physiology, Structure and Function." Trends Microbiology 7, no. 10 (1999): 420–424.

[10] Braithwaite, E., E. Beaugnon, and R. Tournier. "Magnetically Controlled Convection in a Paramagnetic Fluid." Nature 354 (1991): 134–136.

[11] Crabbe, A., et al. "Response of Pseudomonas Aeruginosa PAO1 to Low Shear Modelled Microgravity Involves AlgU Regulation." Environmental Microbiology 12, no. 6 (2010): 1545–1564.

[12] Daly, M.J. "Death by Protein Damage in Irradiated Cells. DNA Repair (Amst), 11, no. 1 (2012): 12–21.

[13] de Vet, S.J., and R. Rutgers. "From Waste to Energy: First Experimental Bacterial Fuel Cells Onboard the International Space Station." Microgravity Science and Technology 19, no. 5–6 (2007): 225–229.

[14] Demain, A.L., and A. Fang. "Secondary Metabolism in Simulated Microgravity." The Chemical Record 1, no. 4 (2001): 333–346.
[15] Dijkstra, C.E., et al. "Diamagnetic Levitation Enhances Growth of Liquid Bacterial Cultures by Increasing Oxygen Availability." Journal of the Royal Society Interface 8, no. 56 (2011): 334–344.
[16] Du, J., and J.M. Gebicki. "Proteins are Major Initial Cell Targets of Hydroxyl Free Radicals." The International Journal of Biochemistry & Cell Biology 36, no. 11 (2004): 2334–2343.
[17] Fredrickson, J.K., et al. "Protein Oxidation: Key to Bacterial Desiccation Resistance?" ISME Journal 2, no. 4 (2008): 393–403.
[18] Goossens, O., et al. "Radiation Dosimetry for Microbial Experiments in the International Space Station Using Different Etched Track and Luminescent Detectors." Radiat Prot Dosimetry 120, no. 1–4 (2006): 433–437.
[19] Hamill, O.P., and B. Martinac. "Molecular Basis of Mechanotransduction in Living Cells." Physiological Review 81, no. 2 (2001): 685–740.
[20] Hammond, T.G., et al. "Mechanical Culture Conditions Effect Gene Expression: Gravity-Induced Changes on the Space Shuttle." Physiological Genomics 3, no. 3 (2000): 163–173.
[21] Hammond, T.G., and J.M. Hammond. "Optimized Suspension Culture: The Rotating-Wall Vessel." American Journal of Physiology 281, no. 1 (2001): F12–F25.
[22] Horswill, A.R., et al. "The Effect of the Chemical, Biological, and Physical Environment on Quorum Sensing in Structured Microbial Communities." Analytical and Bioanalytical Chemistry 387, no. 2 (2007): 371–380.
[23] Hoson, T., et al. "Evaluation of the Three-Dimensional Clinostat as a Simulator of Weightlessness." Planta 203 no. Suppl (1997): S187–S197.
[24] Ingber, D. "How Cells (might) Sense Microgravity." FASEB Journal 13 Suppl (1999): S3–S15.
[25] Ingber, D.E. "Integrins, Tensegrity, and Mechanotransduction." Gravit and Space Biology Bulletin 10, no. 2 (1997): 49–55.
[26] Johanson, K., et al. "Saccharomyces Cerevisiae Gene Expression Changes During Rotating Wall Vessel Suspension Culture." The Journal of Applied Physiology 93, no. 6 (2002): 2171–2180.
[27] Klaus, D.M., P. Todd, and A. Schatz. "Functional Weightlessness During Clinorotation of Cell Suspensions." Advances Space Research 21, no. 8–9 (1998): 1315–1318.

[28] Klaus, D., et al. "Investigation of Space Flight Effects on Escherichia coli and a Proposed Model of Underlying Physical Mechanisms." Microbiology 143, no. Pt 2 (1997): 449–455.

[29] Lapchine, L., et al. "Antibiotic activity in space." Drugs Under Experimental and Clinical Research 12 no. 12 (1986): 933–938.

[30] Leroy, B., et al. "Differential Proteomic Analysis Using Isotope-Coded Protein-Labeling Strategies: Comparison, Improvements and Application to Simulated Microgravity Effect on Cupriavidus Metallidurans CH34." Proteomics 10, no. 12 (2010): 2281–2291.

[31] Mastroleo, F., et al. "Modelled Microgravity Cultivation Modulates N-acylhomoserine Lactone Production in Rhodospirillum Rubrum S1H Independently of Cell Density." Microbiology 159, no. Pt 12 (2013): 2456–2466.

[32] Mastroleo, F., et al. "Experimental Design and Environmental Parameters Affect Rhodospirillum Rubrum S1H Response to Space Flight." ISME Journal 3, no. 12 (2009): 1402–1419.

[33] Mauclaire, L., and M. Egli. "Effect of Simulated Microgravity on Growth and Production of Exopolymeric Substances of Micrococcus Luteus Space and Earth Isolates." FEMS Immunology and Medical Microbiology 59, no. 3 (2010): 350–256.

[34] McPherson, A. "Effects of a Microgravity Environment on the Crystallization of Biological Macromolecules." Microgravity Science and Technology 6, no. 2 (1993): 101–109.

[35] Meloni, M.A., et al. "Cytoskeleton Changes and Impaired Motility of Monocytes at Modelled Low Gravity." Protoplasma 229, no. 2–4 (2006): 243–249.

[36] Nauser, T., W.H. Koppenol, and J.M. Gebicki. "The Kinetics of Oxidation of GSH by Protein Radicals." Biochemical Journal 392, no. Pt 3 (2005): 693–701.

[37] Nickerson, C.A., et al. "Microgravity as a Novel Environmental Signal Affecting Salmonella Enterica Serovar Typhimurium Virulence." Infection and Immunity 68, no. 6 (2000): 3147–3152.

[38] Nickerson, C.A., et al. "Low-Shear Modeled Microgravity: A Global Environmental Regulatory Signal Affecting Bacterial Gene Expression, Physiology, and Pathogenesis." The Journal of Microbiological Methods 54, no. 1 (2003): 1–11.

[39] Nickerson, C.A., et al. "Microbial Responses to Microgravity and Other Low-Shear Environments." Microbiology and Molecular Biology Reviews 68, no. 2 (2004): 345–361.

[40] Novikova, N., et al. “Survey of Environmental Biocontamination on Board the International Space Station.” Research in Microbiology 157, no. 1 (2006): 5–12.

[41] Pivetti, C.D., et al. “Two Families of Mechanosensitive Channel Proteins.” Microbiology and Molecular Biology Reviews 67, no. 1 (2003): 66–85.

[42] Pyle, B., M. Vasques, and R. Aquilina. “The Effect of Microgravity on the Smallest Space Travelers.” Bacterial Physiology and Virulence on Earth and in Microgravity. 2002: National Aeronautics and Space Administration (NASA).

[43] Rea, G., et al. “Ionizing Radiation Impacts Photochemical Quantum Yield and Oxygen Evolution Activity of Photosystem II in Photosynthetic Microorganisms.” International Journal Radiation Biology 84, no. 11 (2008): 867–877.

[44] Schiwon, K., et al. “Comparison of Antibiotic Resistance, Biofilm Formation and Conjugative Transfer of Staphylococcus and Enterococcus Isolates from International Space Station and Antarctic Research Station Concordia. Microbial Ecology 65, no. 3 (2013): 638–651.

[45] Thevenet, D., R. D’Ari, and P. Bouloc. “The SIGNAL Experiment in BIORACK: Escherichia coli in Microgravity.” Journal of Biotechnolgy 47, no. 2–3 (1996): 89–97.

[46] Thomas, W.E., et al. “Bacterial Adhesion to Target Cells Enhanced by Shear Force.” Cell 109, no. 7 (2002): 913–923.

[47] Timmery, S., X. Hu, and J. Mahillon. “Characterization of Bacilli Isolated from the Confined Environments of the Antarctic Concordia Station and the International Space Station.” Astrobiology 11, no. 4 (2011): 323–334.

[48] Tixador, R., et al. “Study of Minimal Inhibitory Concentration of Antibiotics on Bacteria Cultivated In Vitro in Space (Cytos 2 experiment).” Aviation, Space, and Environmental Medicine 56, no. 8 (1985): 748–751.

[49] Tucker, D.L., et al. “Characterization of Escherichia coli MG1655 Grown in a Low-Shear Modeled Microgravity Environment.” BMC Microbiology 7 (2007): 15.

[50] Valles, Jr. J.M., et al. “Stable Magnetic Field Gradient Levitation of Xenopus Laevis: Toward Low-Gravity Simulation.” Biophysical Journal 73, no. 2 (1997): 1130–1133.

[51] Van Houdt, R., et al. “Evaluation of the Airborne Bacterial Population in the Periodically Confined Antarctic Base Concordia.” Microbial Ecology 57, no. 4 (2009): 640–648.

[52] Vukanti, R., E. Mintz, and L. Leff. "Changes in Gene Expression of E. coli Under Conditionss of Modeled Reduced Gravity." Microgravity Science and Technology 20 (2008): 41–57.

[53] Walther, I., et al. "Simulated Microgravity Inhibits the Genetic Expression of Interleukin-2 and its Receptor in Mitogen-Activated T Lymphocytes." FEBS Letter 436, no. 1 (1998): 115–118.

[54] Wilson, J.W., et al. "Space Flight Alters Bacterial Gene Expression and Virulence and Reveals a Role for Global Regulator Hfq." Proceedings of the National Acadamic Science of the United States of America 104, no. 41 (2007): 16299–16304.

[55] Wilson, J.W., et al. "Low-Shear Modeled Microgravity Alters the Salmonella Enterica Serovar Typhimurium Stress Response in an RpoS-Independent Manner." Applied and Environmental Microbiology 68, no. 11 (2002): 5408–5016.

23

Gravitational Cell Biology

Cora S. Thiel and Oliver Ullrich

University of Zurich, Zurich, Switzerland

23.1 Gravitational Cell Biology

The evolution of life on Earth has been subject to a range of influences, whereas gravity was the only constant and universal force during all times of Earth's history. Since the last decades, a lot of evidence has been obtained suggesting that the function of mammalian cells and of small unicellular organisms is different under conditions of microgravity. Consequently, the question arose of how normal gravity may play a role in "normal" cellular function and if gravity may provide important signals for the cell. It is a common method to investigate the effects of forces by using systems in which these forces can be eliminated. In the case of gravity, this means that experimental environments have to be created where no or only minor gravitational forces prevail. A variety of platforms exist where these conditions can be achieved. Experiments can be performed under microgravity conditions of different length varying between seconds (drop tower or parabolic flights), minutes (sounding rockets) until up to permanent microgravity (ISS).

23.2 Studies Under Simulated Microgravity

However, the accessibility of these platforms is limited and cannot fully cover the needs of the research community. Therefore, platforms providing access to simulated microgravity are of great interest because they offer almost unlimited experimental capacity. The most frequently used ground-based facilities for these purposes (described in detail in Chapters 8 and 14) are the 2D clinostat, the random positioning machine (RPM), the rotating wall vessel (RWV), and the diamagnetic levitation [1]. These devices were not only used for experiments with unicellular organisms such as Euglena, Paramecium, or Loxodes, but also for experiments with whole plants, animals, or plant and animal cell cultures. Simulated gravity studies with protists showed valuable results concerning their gravitaxis and gravikinesis [2, 3]. A comparison of the ground-based facilities used in experiments with the protists Euglena and Paramecium showed that simulated microgravity conditions generated by the fast rotating 2D clinostat are well suited, whereas the organisms experience a change in positive and negative acceleration forces when they are exposed to simulated microgravity on the RPM [1]. Experiments using magnetic levitation are less recommended, because of the strong effects of the magnetic field acting on these two types of protists [1, 4, 5].

23.3 Effects of Simulated Microgravity on Algae, Plant Cells, and Whole Plants

Simulated gravity is also used in experiments with algae, plant cell cultures, and whole plants. In Chara, the statolith-based gravity-sensing system in the rhizoids has been intensively investigated by 2D and 3D clinorotation as well as by magnetic levitation [6–8]. The comparison of the experimental outcomes showed that when Chara is used as a study object, the fast-rotating 2D clinostat is the preferred equipment to be applied for simulated microgravity, followed by the 3-D clinostat and the random positioning machine. In contrast, magnetic fields were not sufficient to generate the required simulated microgravity conditions [1].

Arabidopsis thaliana is a frequently used study object because of its excellent characterization on the molecular level. Gene expression analysis performed with whole plants on a 3D clinostat described the identification of new simulated microgravity stress-induced transcription factors [9]. Investigations performed with Arabidopsis cell cultures on 2D and 3D clinostats as well as under magnetic levitation showed altered gravity-related signaling cascades in the undifferentiated cultured cells [10].

23.4 Mammalian Cells in Simulated Microgravity

Mammalian cell cultures are frequently employed to study the direct effects of microgravity on the fundamental processes in the cells of our body. Cell lines, especially with fibroblast characteristics, are often used because of the easy handling and experiment preparations. It could be shown that under real and simulated microgravity conditions (2D clinostat or RPM), the epidermal growth factor (EGF)-induced signaling pathway is affected and that cells changed their morphology to a roundish shape due to modifications affecting the cytoskeleton [11–13]. New experimental designs allow also for the use of the RPM to expose cells or even small organisms to simulated partial gravity [presented by Dutch Space (Leiden, The Netherlands) at the European Low Gravity Research Association in Vatican City in 2013 and in reference [14]].

Numerous comparative studies between real and simulated microgravity have also been performed with semi-adherent or non-adherent cells of the immune system. Semi-adherent NR8383 macrophages show a similar behavior under real and simulated microgravity conditions. The macrophageal oxidative burst reaction is one of the key functions in innate immune response. The associated phagocytosis-mediated reactive oxygen species (ROS) production is rapidly and reversely decreased upon reduced gravitation. The direct comparison between ground based and investigations under real microgravity during parabolic flights using a 2D clinostat combined with a one-axis clinostat fitted with a photo multiplier tube showed for both experimental setups a pronounced reduction in the ROS production and therefore of the reactivity of the cells [15]. Comparative studies have been also performed on U937 cells, a human myelomonocytic cell line. For both, real and simulated microgravity, a decreased proliferation rate was observed [16, 17]. Furthermore, a reduced locomotion was identified in monocytes most likely due to cytoskeletal modifications such as a reduced density of the actin filaments and impaired ß-tubulin architecture [18, 19]. T lymphocytes also strongly react to microgravity: They show an activation failure upon stimulation with ConA (for a review, see [20]). This effect was reproducible under simulated microgravity conditions in the RPM [21, 22]. In line with these results, stimulation of Jurkat T cells with PMA resulted in an increased expression of the cell cycle-regulating protein p21Waf1/Cip1 after 15min of 2D clinorotation. Real-time PCR experiments confirmed these results on gene expression level [23]. Importantly, differential gene expression studies from RPM experiments revealed that impaired induction of early genes regulated primarily by transcription factors NF-kappaB, CREB, ELK, AP-1, and STAT after T-cell receptor cross-linking contributed to T-cell dysfunction in altered gravity, associated with down-regulation of the PKA pathway [24].

Taken together, ground-based facilities for simulated microgravity are of great value and should be frequently used for preparation and/or validation of real microgravity experiments, where time and sample numbers are often limited. Several studies could show that 2D clinostats and the random positioning machine (RPM) are suitable to perform these ground-based simulation experiments. However, during the experiment design phase, special attention shall be paid on choosing the most suitable system.

References

[1] Herranz, R., R. Anken, J. Boonstra, M. Braun, P.C.M. Christianen, M. de Geest, J. Hauslage, R. Hilbig, R. J.A. Hill, M. Lebert, F. J. Medina, N. Vagt, O. Ullrich, J. J.W.A. van Loon, and R. Hemmersbach. “Ground-Based Facilities for Simulation of Microgravity: Organism-Specific Recommendations for Their Use, and Recommended Terminology.” *Astrobiology* 13 (2013): 1–17.

[2] Hemmersbach, R., and D.-P. Haeder. “Graviresponses of Certain Ciliates and Flagellates.” *FASEB Journal* 13 (1999): S69–S75.

[3] Hemmersbach, R., S.M. Strauch, D. Seibt, M. Schuber. “Comparative Studies on Gravisensitive Protists on Ground (2D and 3D Clinostats) and in Microgravity. *Microgravity Science and Technology* XVIII-3/4 (2006): 257–259.

[4] Guevorkian, K., and J.M. Valles, Jr. “Aligning Paramecium Caudatum with Static Magnetic Fields.” Biophysical Journal 90 (2006): 3004–3011.

[5] Guevorkian, K., J.M. Valles, Jr. “Swimming Paramecium in Magnetically Simulated Enhanced, Reduced, and Inverted Gravity Environments.” *Proceedings of the National Academy of Sciences of the United States of America* 103 (2006): 13051–13056.

[6] Cai, W., M. Braun, and A. Sievers. “Displacement of statoliths in Chara rhizoids during horizontal rotation on clinostats.” *Shi Yan Sheng Wu Xue Bao* 30 (1997): 147–155.

[7] Braun, M., B. Buchen, and A. Sievers. “Actomyosin-Mediated Statolith Positioning in Gravisensing Plant Cells Studied in Microgravity.” *The Journal of Plant Growth Regulation* 21 (2002): 137–145.

[8] Limbach, C., J. Hauslage, C. Schafer, and M. Braun. “How to Activate a Plant Gravireceptor. Early Mechanisms of Gravity Sensing Studied in Characean Rhizoids During Parabolic Flights.” *Plant Physiology* 139 (2005): 1030–1040.

[9] Soh, H., Y. Choi, T.-K. Lee, U.-D. Yeo, K. Han, C. Auh, and S. Lee. “Identification of Unique Cis-Element Pattern on Simulated Microgravity Treated Arabidopsis by in Silico and Gene Expression.” *Advances in Space Research* 50 (2012): 397–407.

[10] Babbick, M., C. Dijkstra, O.J. Larkin, P. Anthony, M.R. Davey, J.B. Power, K.C. Lowe, M. Cogoli-Greuter, and R. Hampp. “Expression

of Transcription Factors After Short-Term Exposure of Arabidopsis Thaliana Cell Cultures to Hypergravity and Simulated Microgravity (2-D/3-D clinorotation, magnetic levitation). *Advances in Space Research* 39 (2007): 1182–1189.

[11] Boonstra. J. "Growth Factor-Induced Signal Transduction in Adherent Mammalian Cells is Sensitive to Gravity." *FASEB Journal* 13 (1999): S35–S42.

[12] Moes, M., J. Boonstra, and E. Regan-Klapisz. "Novel Role of cPLA(2)Alpha in Membrane and Actin Dynamics." *Cellular and Molecular Life Sciences* 67 (2010): 1547–1557.

[13] Pietsch, J., X. Ma, M. Wehland, G. Aleshcheva, A. Schwarzwälder, J. Segerer, M. Birlem, A. Horn, J. Bauer, M. Infanger, and D. Grimm. "Spheroid Formation of Human Thyroid Cancer Cells in an Automated Culturing System During the Shenzhou-8 Space mission." *Biomaterials* 34, no. 31 (2013): 7694–7705.

[14] Wuest, S.L., S. Richard, S. Kopp, D. Grimm, and M. Egli. "Simulated Microgravity: Critical Review on the Use of Random Positioning Machines for Mammalian Cell Culture." *Bio Med Research International*, 2015, Article ID 971474, (2014): 8 pp.

[15] Adrian, A., K. Schoppmann, J. Sromicki, S. Brungs, M. von der Wiesche, B. Hock, W. Kolanus, R. Hemmersbach, and O. Ullrich. "The Oxidative Burst Reaction in Mammalian Cells Depends on Gravity." *Cell Communication and Signaling* 11 (2013): 98.

[16] Hatton, J. P., F. Gaubert, M. L. Lewis, Y. Darsel, P. Ohlmann, J. P. Cazenave, and D. Schmitt. "The Kinetics of Translocation and Cellular Quantity of Protein Kinase C in Human Leukocytes are Modified During Spaceflight." *FASEB Journal* 13, (supplement) (1999): S23–S33.

[17] Villa, A., S. Versari, J. A.Maier, S. Bradamante. "Cell Behavior in Simulated Microgravity: A Comparison of Results Obtained with RWV and RPM." *Gravitational and Space Biology Bulletin* 18 no. 2 (2005): 89–90.

[18] Meloni, M.A., G. Galleri, P. Pippia, and M. Cogoli-Greuter. "Cytoskeleton Changes and Impaired Motility of Monocytes at Modelled Low Gravity." *Protoplasma* 229, no. 2–4 (2006): 243–249.

[19] Meloni, M.A., G. Galleri, G. Pani, A. Saba, P. Pippia, and M. Cogoli-Greuter. "Space Flight Affects Motility and Cytoskeletal Structures in Human Monocyte Cell Line J-111." *Cytoskeleton* 68, no. 2 (2011): 125–137.

[20] Cogoli-Greuter, M. "The Lymphocyte Story—An Overview of Selected Highlights on the In Vitro Activation of Human Lymphocytes in Space." *Microgravity Science and Technology*, 25, no. 6 (2014): 343–352.
[21] Walther, I., P. Pippia, M. A. Meloni, F. Turrini, F. Mannu, and A. Cogoli. "Simulated Microgravity Inhibits the Genetic Expression of Interleukin-2 and its Receptor in Mitogen-Activated T Lymphocytes." *FEBS Letters* 436, no. 1 (1998): 115–118.
[22] Schwarzenberg, M., P. Pippia, M.A. Meloni, G. Cossu, M. Cogoli-Greuter, and A. Cogoli. "Signal Transduction in T Lymphocytes—A Comparison of the Data from Space, the Free Fall Machine and the Random Positioning Machine." *Advances in Space Research* 24, no. 6 (1999), 793–800.
[23] Thiel, C.S., K. Paulsen, G. Bradacs, K. Lust, S. Tauber, C. Dumrese, A. Hilliger, K. Schoppmann, J. Biskup, N. Goelz,, C. Sang,U. Ziegler, K. H. Grote, F. Zipp, F. Zhuang, F. Engelmann., R. Hemmersbach, A. Cogoli, and O. Ullrich. "Rapid Alterations of Cell Cycle Control Proteins in Human T Lymphocytes in Microgravity." *Cell Communication and Signaling* 10 (2012): 1.
[24] Boonyaratanakornkit, J.B., A. Cogoli, C.F. Li, T. Schopper, P. Pippia, G. Galleri, M.A. Meloni, M. Hughes-Fulford. "Key Gravity-Sensitive Signaling Pathways Drive T Cell Activation." *FASEB Journal* 19, no. 14 (2005): 2020–2.

24

Growing Plants under Generated Extra-Terrestrial Environments: Effects of Altered Gravity and Radiation

F. Javier Medina[1], Raúl Herranz[1], Carmen Arena[2], Giovanna Aronne[3] and Veronica De Micco[3]

[1]Centro de Investigaciones Biológicas (CSIC) Madrid, Spain
[2]University of Naples Federico II, Department of Biology, Naples, Italy
[3]University of Naples Federico II, Department of Agricultural and Food Sciences, Portici (Naples), Italy

24.1 Introduction: Plants and Space Exploration

The coming enterprises of space exploration will surely require the cultivation of plants, not only as part of Bioregenerative Life Support Systems (BLSS), but also as a source of psychological well-being for space travelers. Considering that gravity plays a unique role in the configuration of a normal plant developmental pattern, only comparable with the effect of light, weightlessness is a major stress condition that should be fully understood.

More than 50 years have passed since the first experiments performed in Space demonstrated that seed germination and plant growth can happen under conditions of altered gravity. Although with sometimes-contrasting results, data from experiments in real or simulated weightlessness indicate that microgravity itself does not prevent plant growth and reproduction; besides, the *seed-to-seed* cycle can be accomplished for several consecutive generations [1]. Indeed, during the evolution, being non-migrating organisms, plants have developed a high plasticity to adapt to changing environmental conditions. Such a high plasticity is the basis for coping with the limiting factors of extra-terrestrial environments. Morpho-functional changes due to microgravity at different plant levels have been attributed to the direct effect of microgravity on common metabolic pathways and on subcellular processes. Moreover, the interaction between microgravity and other factors, either Space-related (e.g., high levels of radiation) or distinctive of BLSS (e.g., atmospheric composition of the closed systems), can be more effective in determining growth and reproductive aberrations than the sole altered gravity [2].

Such environmental factors include quantity and quality of light, atmosphere composition of the pressurized modules, availability of water and nutrients, which may occur at suboptimal levels and would act synergistically to gravity alteration, enhancing the effects of gravitational stress. The mechanism of this synergistic effect in microgravity environments is only partially understood although a number of model systems have been successfully exposed to both real spaceflight and simulated microgravity conditions [reviewed by 3]. In fact, those effects are even more obscure under fractional gravity conditions, like the ones on Mars (0.37 g) or the Moon (0.17 g) surface. In fractional gravity conditions, some of the mechanisms that are disturbed in microgravity could still be affected (e.g., reduced graviresistance in cell walls) but others would remain unaltered (e.g., still working gravitropic responses at those gravity levels). It has been hypothesized that the two components of the gravity vector, namely direction and magnitude, could be sensed differentially in these conditions [4].

In terms of research, to generate on Earth a reduced-gravity environment is also not straightforward and requires a comparable compromise between good quality of simulation and side effects of the technology used for microgravity simulation. For instance, magnetic levitation of samples provides a very stable partial-gravity environment, but adds a new layer of complexity due to the presence of high magnetic fields in the equation. Research on plant gravitropic responses has also profited by experiments in hypergravity conditions simulated through centrifuges [5–7]. It should be reminded here the usefulness of hypergravity research in other biological systems to learn about diseases as osteoporosis or sport training under overloading environments.

The accurate identification of stress factors, and the strategy followed by plants in their adaptation to an extra-terrestrial environment, will be necessary for achieving a successful cultivation of plants on board of spaceships and, in general, outside the Earth environment, which is essential to make manned space exploration possible.

In this section, we report an overview of the current knowledge about plant's responses to altered gravity from molecular to the whole organism level. In the light of long-lasting experimentation in space, or in simulated space conditions, the influence of ionizing radiation and possible interferences of other environmental factors acting at suboptimal levels in ecologically closed systems are also considered as additional stressors for the achievement of efficient plant growth.

24.2 Cellular and Molecular Aspects of the Gravity Perception and Response in Real and Simulated Microgravity

24.2.1 Gravity Perception in Plant Roots: Gravitropism

Among environmental factors influencing plant growth, development, survival, and evolution, gravity is characterized by its permanent and constant presence on Earth. It is long-time known that plants are sensitive to the presence of the gravity vector which drives plant growth direction by a phenomenon called gravitropism. The process is triggered by the displacement of amyloplasts (statoliths) in the columella root cap cells [8].

The mechanism of plant root gravitropism involves, first, the transformation of the mechanical signal into a biochemical signal, and then, the transduction of this signal from the site of sensing to the site of response; this process is mediated by the phytohormone auxin and its polar transport throughout the root length [9, 10] (Figure 24.1). This mechanism has been

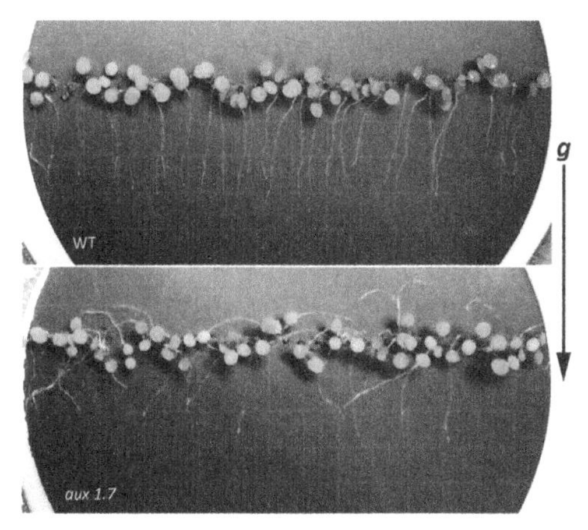

Figure 24.1 Seedlings of *Arabidopsis thaliana*. The upper image shows the wild type and the lower image the agravitropic aux 1.7 mutant. Seedlings of the wild type show conspicuous gravitropic behavior, with the roots aligned in the direction of the gravity vector; however, aux 1.7 mutant seedlings show evident alterations of gravitropism with roots growing in random directions.

mainly elucidated by experiments consisting of changing the orientation of growing seedlings. It was shown that a short time after seedling re-orientation, the root changes the direction of its growth according to the gravity vector. Root bending associated with the change in growth direction is caused by the alteration of the auxin polar transport, leading to the lateral re-distribution of auxin in the distal regions of the root, which eventually results in a different rate of elongation of cells located at both sides of the root in a specific zone (elongation zone) [11]. This morphological event is accompanied by a substantial reorganization of gene expression [12].

An important source of information on the mechanisms of gravitropism comes from studies performed in the absence of gravity. On Earth, real weightlessness is obtained by free fall; however, values of the gravity vector near zero (lower than 10^{-6} g) occur in long-duration missions, in spaceships orbiting the Earth, such as the International Space Station (ISS), or, previously, the Space Shuttle.

In early experiments in the Spacelab, it was shown that lentil root cell statoliths did not distribute at random, but they accumulated in the proximal region of columella cells [13]. In fact, statoliths are attached to actin filaments by means of myosin, and this greatly affects their displacement within the cell [14]. Otherwise, the use of starchless *Arabidopsis* mutants in spaceflight experiments provided support to the starch-statolith model for gravity sensing [15].

Space experiments have provided support to the role of auxin polar transport as a fundamental mediator of the transduction of the gravity signal. Seedlings grown in space showed alterations in the auxin polar transport [16], and genes related to this process have been found to be differentially regulated in spaceflight- or parabolic-flight-grown samples [17–19].

The intrinsic constraints of space experimentation, mostly the limitations of the general access to spaceflights, induced the development of ground-based devices capable of a reliable simulation of microgravity conditions. For this purpose, devices capable of counteracting the perception of the Earth gravity vector by plants, called clinostats, were designed and constructed. The first classical clinostats were introduced by Sachs in 1879 and allowed to work with simulated microgravity on Earth [reviewed by 3]. It is important to emphasize, however, that these devices, including the most modern ones, such as the random positioning machine (RPM), do not suppress the gravity vector, but they only act at the level of the mechanism by which living beings perceive it. In addition, recent technologies such as diamagnetic levitation can be used to produce a weightlessness environment for plants. This technology takes advance of the major presence of diamagnetic materials in organisms (mostly water): by producing a high gradient of magnetic field into the small volume of the sample container in the magnet bore (a tube of around 5 cc), a strong levitation force is produced, which is capable of counteracting gravity. Although this technology has the advance of producing a continuous (not averaged) effect, even detectable at the molecular level, it has the disadvantage of adding side effects, mostly the magnetic field itself, into the equation. On the other hand, it allows performing several partial *g* experiments (in the gravity range between 0 and 2 g) at the same time and in the same environment [3].

Ground-based facilities for microgravity simulation have demonstrated their usefulness in gravity-related research. In some cases, differences between the statolith position in clinostats and in space have been reported [13], but, in general, results are largely comparable [20].

24.2.2 Effects on Cell Growth and Proliferation

Altered gravity produces changes in plant development, which are closely associated with modifications in the activities of meristems, specifically cell growth and cell proliferation/cell cycle. All adult plants contain meristematic tissues, composed by populations of undifferentiated, highly proliferating cells, capable of forming any specialized tissue at any time in the life span of the plant. Indeed, plant development greatly depends on the balance between cell proliferation and cell differentiation in the meristems; these phenomena are controlled, in turn, by the auxin [21]. Results obtained in space experiments showed altered cell proliferation rate in both lentil [22] and *Arabidopsis* [23]. Under microgravity conditions, cell proliferation and cell growth appear uncoupled, losing their coordinated progress which is called "meristematic competence" and is the main feature of these cells under normal ground gravity conditions [9, 23, 24].

It should be pointed out that spaceflight not only causes gravitational stress, but it also involves additional environmental modifications with respect to the Earth conditions, such as the confinement, the lack of gas convection, and the cosmic radiation, which may be sensed by plants, causing additional stresses [25]. For this reason, the results obtained in real microgravity have to be sustained by studies carried out on ground, using the available facilities for microgravity simulation [3] in combination with some of the stressors that could be present in extraterrestrial habitats.

24.2.3 Effects of Gravity Alteration on Gene Expression

A few studies on gene expression have been carried out in real or simulated microgravity, by performing an overall transcriptomic analysis of seedlings or solid cell cultures (calluses), to search for gravity-regulated genes. Experiments included reorientation of seedlings [12], cell cultures grown in ground-based facilities for microgravity simulation [26, 27], and growth of seedlings during parabolic flights [19] or in spaceflight conditions [17, 18]. These studies have revealed a complex response of plants to gravity alteration, including changes in the expression of genes involved in general responses to other stresses, such as drought, cold, light, and biotic stimulation by pathogens. Heat shock-related genes, cell wall remodeling, and cell expansion genes, as well as genes involved in oxidative burst and plant defense, are outstanding examples of transcriptomic alterations shared by different kinds of plant stresses, including the gravitational one, and commonly reported in different experiments. However, in most studies, the existence of a specific response to gravity alteration was suggested or postulated. Specific alterations affect genes involved in the gravitropic response, such as those related to hormone signaling, particularly auxin [17–19], and also a number of genes of unknown function. Furthermore, there is some specificity in the process of adaptation to spaceflight environment by different organs of the plant, which engages different genes, even though the general strategy of response is shared in all cases [18]. An additional specific feature of gravity, as a permanent component of the environment, is that it is capable of altering the perception and response of living organisms to other environmental factors [26, 28].

24.3 Morpho-Functional Aspects of the Plant Response to Real and Simulated Microgravity Environments

24.3.1 From Cell Metabolism to Organogenesis

Organogenesis is a complex phenomenon requiring a precise coordination of cell proliferation, enlargement, and differentiation, which are regulated by gene expression and signaling biologically active molecules. The mechanisms linking cell division and the whole plant development have not been completely understood yet [29]. As reported above, microgravity surely affects cell proliferation which is the first step for plant growth; in addition, cell enlargement and differentiation can be modified in altered gravity conditions. Most of the knowledge about the effect of microgravity on morphogenesis in higher plants derives from experiments where seedlings, being the smallest yet complete form of a plant, have been used as models due to volume and time constraints. The whole plant life cycle has been investigated only in a few plant species characterized by small size, such as *Arabidopsis* spp. and *Brassica* spp. [2]. Indeed, vegetal systems characterized by reduced size better match with volume constraints in the experimental facilities in space and do not encounter bending problems during clinorotation. The interest in studying seedlings also derives from their high nutritional value which makes them suitable candidates to reintegrate the crew's diet with fresh food [30].

Morpho-anatomical changes detected in plants growing under real or simulated microgravity have been often considered primarily as the consequence of altered cytoskeleton organization, increased production of reactive oxygen species (ROS), and modifications in starch and phenylpropanoid metabolism. Cortical microtubule organization has been demonstrated to be affected by reduced or increased gravity levels in both gravisensing cells and in cells not specialized to gravity perception in isolated protoplasts, root, and other organs [31–33]. The alteration of spatial organization of cytoskeleton microtubules has influence on cell growth because it not only regulates the cytoplasm expansion, but also affects cellulose microfibrils orientation during cell wall development, which in turn controls protoplast expansion [34, 35]. Increased or decreased cell enlargement, sometimes ascribed to the accumulation of ethylene in the experiment containers during spaceflight, have been related to changes in seedling size [36]. The size of roots or hypocotyls was found to increase, decrease, or remain unchanged in seedlings of various species exposed to clinorotation or real microgravity (see, for example, [37, 38]). Relatively few studies have focused on leaf development in space also with contrasting results. Stutte et al. [39] showed that the minimal space-induced modifications in leaf structure (e.g., reduced thickness, more dense mesophyll, and altered chloroplast morphology accompanied by unchanged contents in starch, soluble sugars, and lignin) of *Triticum aestivum* did not cause any physiological changes. These authors suggested that severe alterations in the (ultra)structure of tissues and organelles, found in early spaceflight experiments, were caused by an inadequate environmental control in the experimental chambers.

24.3.2 Indirect Effects of Altered Gravity to Photosynthesis

The maintenance of high photosynthetic rate is needed to optimize O_2 production and CO_2 removal in BLSS. Experiments conducted on dwarf wheat on the ISS showed that microgravity does not alter the development of the photosynthetic apparatus and the efficiency of photosynthesis [39]. While no changes in chloroplasts membranes were found in flown *Arabidopsis thaliana*, changes in the thylakoid structure occurred in *T. aestivum* and *B. rapa* (reviewed in [31]). Moreover, a decrease in the electron transport rate of photosystem I (PSI), measured in *B. rapa* suggested the major susceptibility of the PSI in comparison with PSII. It has been recently highlighted that some of the structural and physiological changes of the photosynthetic apparatus in space should not be considered as the result of a direct effect of microgravity, but as the consequence of starch accumulation due to the delayed long-distance transport of photosynthetic metabolites [31, 40].

24.3.3 Constraints in the Achievement of the Seed-to-Seed Cycle in Altered Gravity

The reproductive cycle is characterized by a succession of highly specific phases where mitosis and meiosis, as well as cell enlargement, follow one another in a few specialized cells. Whatever the factor affecting cell cycles and cell growth, it can also affect one or more reproductive phases, thus ultimately endangering seed set.

There is a common agreement that the reproductive failure claimed in the early experiments performed in space was triggered by inefficient environmental control in the growth chambers resulting in too high humidity, accumulation of ethylene, and insufficient carbon supply (reviewed in [2]). Problems in ventilation have been considered responsible for the interruption of the plant life cycle in different phases, such as the transition from vegetative to reproductive stage, microsporogenesis, female gametophyte formation, anther dehiscence, pollination, and embryo development. Recent advances in space-related technologies allowed better control of growth chambers, thus facilitating the completion of the *seed-to-seed* cycle in various model species [41, 42]. However, there is evidence that the quality of seeds produced in microgravity is not optimal: seeds of *B. rapa* formed in microgravity were characterized by smaller size and different composition of the reserves compared to controls, probably because microgravity affects the microenvironment inside the silique [42]. Finally, it cannot be disregarded that some reproductive phases are strictly controlled by the coordinate activity of intracellular components, particularly cytoskeleton, which is affected by microgravity. For example, pollen tube development in simulated microgravity is subjected to alterations which could reduce its capacity of fertilization [43]. To conclude, alterations in different reproductive phases, due to direct or indirect effects of microgravity, can either completely interrupt the reproductive cycle or determine progressive lowering of the reproductive success.

24.4 Plant Response to Real or Ground-Generated Ionizing Radiation

24.4.1 Variability of Plant Response to Ionizing Radiation

Space radiation represents a major barrier to human exploration of the solar system because of the biological effects of high-energy and charge (HZE) nuclei [44]. The degree of damage is associated with different radiation types. High-linear energy transfer (LET)-ionizing radiations (e.g., protons and heavy ions) have lower penetrating capability but are more dangerous than low-LET ones (e.g. X- and gamma-rays) for both plant and mammalian cells [45–47].

The current knowledge about the biological effects of radiation on photosynthetic organisms comes from experiments performed in long- and short-duration space missions. Additional sources of information derive from ground-based experiments conducted by means of accelerators testing different ions and energy ranges on diverse species. Despite the high variability in the results, which often makes data hardly comparable, a common view is that ionizing radiation can have stimulatory effects at very low doses, harmful consequences at middle levels, and detrimental outcomes at high doses on plant development [46]. The severity of the effects is dependent upon several factors including the species, cultivar, phenological phase, and genome organization [48].

24.4.2 Effects of Ionizing Radiation at Genetic, Structural, and Physiological Levels

Radiation-induced alterations are generally ascribed to the interaction of the radiation itself with atoms and molecules, which causes the production of ROS [49].

At genomic level, the structural and functional changes in the DNA are responsible for most of the damage expressed after the exposure to ionizing radiations. The nature of DNA modifications is variable and includes single-base alterations, base substitutions, base deletions, chromosomal aberrations, and epigenetic modifications. Generally, chronically irradiated samples show a higher genomic instability compared to acutely exposed samples [50].

Radiation-induced morpho-functional changes can be either the phenotypical expression of DNA aberrations or the consequence of the structural damage of tissues. The main alterations include reduced germination, lethality, loss of apical dominance, induction of dwarf growth, altered leaf anatomy, sterility, and accelerated senescence [46, 51]. The degradation of cell wall materials determines tissue softening, due to the dissolution of middle lamellae, and the increase in seed germination, ascribed to increased water absorption due to augmented porosity of seed teguments [52, 53].

Among processes regulating plant life, photosynthesis is one of the most sensitive to radiation. Several components of the photosynthetic machinery may be altered: light-harvesting complexes, electron transport carriers, and enzymes of the carbon reduction cycle [51, 54]. The severity of damage depends on dose; the loss of PSII functionality, and the generation of ROS throughout the cell are observed at very high doses.

The exposure to low doses of ionizing radiation may induce radioresistance [55]. Plants are very radioresistant compared to animals. At cellular level, radioresistance is achieved through the occurrence of mechanical and chemical barriers (e.g., specialized and thickened cell walls, cuticle, pubescence, and phenolic compounds), as well as by increasing the level of ploidy [45, 56, 57]. The chronic exposure to low doses of ionizing radiation also leads to significant differences in the expression of radical scavenging enzymes and DNA-repair genes as well as to an increase in the activity of several antioxidant enzymes in plant tissues [55].

24.5 Conclusions—Living in a BLSS in Space: An Attainable Challenge

More than 50 years of experiments with higher plants in space suggest that almost all developmental and reproductive phases can be successfully overcome, ultimately leading to accomplish the *seed-to-seed* cycle for successive generations. However, possible perturbations can happen at various levels, from molecular to organism, thus lowering the efficiency of growth and reproduction processes. Such perturbations can be ascribed to the interaction between space factors and other environmental factors not acting at optimal levels.

The incessant development of new agro-technologies is promising for the development of BLSS with adequate environmental control to reduce the effect of multiple stressors on plant growth. Ground-based research aiming to the development of BLSS heads for soilless systems to minimize other potential stresses during plant cultivation [58]. However, in space, it is necessary to minimize the use of resources (e.g., lighting, electric power, water, and nutrients consumption). Consequently, a compromise needs to be reached to minimize external supply of resources, still without overcoming the threshold over which suboptimal levels of environmental factors may worsen the effects of microgravity and other space factors on plants. While new agro-technologies progress, the ground-based research with facilities simulating weightlessness, ionizing radiation, and confined environments also needs to advance. Such studies are necessary to elucidate the plant's sensitivity to space factors in the sight of wider space experimentation.

References

[1] Sycheva, V.N., M.A. Levinskikha, S.A. Gostimskyb, G.E. Binghamc, and I.G. Podolskya. "Spaceflight Effects on Consecutive Generations of Peas Grown Onboard the Russian Segment of the International Space Station." *Acta Astronautica* 60 (2007): 426–432.

[2] De Micco, V., S. De Pascale, R. Paradiso, and G. Aronne. "Microgravity Effects on Different Stages of Higher Plant Life Cycle and Completion of the Seed-to-Seed Cycle." *Plant Biology (Stuttg)* 16 Suppl 1 (2014): 31–38.

[3] Herranz, R., R. Anken, J. Boonstra, M. Braun, P.C.M. Christianen, M.D. Geest, J. Hauslage, R. Hilbig, R.J.A. Hill, M. Lebert, F.J. Medina, N. Vagt, O. Ullrich, J.J.W.A. van Loon, and R. Hemmersbach. "Ground-Based Facilities for Simulation of Microgravity: Organism-Specific Recommendations for their Use, and Recommended Terminology." *Astrobiology* 13 (2013): 1–17.

[4] Herranz, R., and F.J. Medina. "Cell Proliferation and Plant Development Under Novel Altered Gravity Environments." *Plant Biology* 16, Suppl. 1 (2014): 23–30.

[5] Musgrave, M.E., A. Kuang, J. Allen, and J.J. van Loon. "Hypergravity Prevents seed Production in Arabidopsis by Disrupting Pollen Tube Growth." *Planta* 230, vol. 5 (2009): 863–870.

[6] Soga, K. "Perception Mechanism of Gravistimuli in Gravity Resistance Responses of Plants." *Biolgical Sciences in Space* 18, vol. 3 (2004): 92–93.

[7] Tamaoki, D., I. Karahara, T. Nishiuchi, T. Wakasugi, K. Yamada, and S. Kamisaka. "Effects of Hypergravity Stimulus on Global Gene

Expression During Reproductive Growth in Arabidopsis." *Plant Biology (Stuttg)* 16, Suppl 1 (2014): 179–86.

[8] Baldwin, K.L., A.K. Strohm, and P.H. Masson. "Gravity Sensing and Signal Transduction in Vascular Plant Primary Roots." *American Journal of Botany* 100, no. 1 (2013): 126–142.

[9] Medina, F.J., and R. Herranz. "Microgravity Environment Uncouples Cell Growth and Cell Proliferation in Root Meristematic Cells: The Mediator Role of Auxin." *Plant Signaling & Behavior* 5, no. 2 (2010): 176–179.

[10] Teale, W.D., I.A. Paponov, and K. Palme. "Auxin in Action: Signalling, Transport and the Control of Plant Growth and Development." *Nature Reviews Molecular Cell Biology* 7, no. 11 (2006): 847–859.

[11] Friml J., J. Wisniewska, E. Benkova, K. Mendgen, and K. Palme. "Lateral Relocation of Auxin Efflux Regulator PIN3 Mediates Tropism in Arabidopsis." *Nature* 415, no. 6873 (2002): 806–809.

[12] Moseyko, N., T. Zhu, H.S. Chang, X. Wang, and L.J. Feldman. "Transcription Profiling of the Early Gravitropic Response in Arabidopsis Using High-Density Oligonucleotide Probe Microarrays." *Plant Physiology* 130 (2002): 720–728.

[13] Perbal, G., D. Driss-Ecole, J. Rutin, and G. Salle. "Graviperception of Lentil Roots Grown in Space (Spacelab D1 Mission)." *Physiologia Plantarum* 70 (1987): 119–126.

[14] Perbal, G., and D. Driss-Ecole. "Contributions of Space Experiments to the Study of Gravitropism." *Journal of Plant Growth Regulation* 21 (2002): 156–165.

[15] Kiss, J.Z., W.J. Katembe, and R.E. Edelmann. "Gravitropism and Development of Wild-Type and Starch-Deficient mutants of Arabidopsis During Spaceflight. *Physiologia Plantarum* 102 (1998): 493–502.

[16] Ueda, J., K. Miyamoto, T. Yuda, T. Hoshino, S. Fujii, C. Mukai, S. Kamigaichi, S. Aizawa, I. Yoshizaki, T. Shimazu, and K. Fukui. "STS-95 Space Experiment for Plant Growth and Development, and Auxin Polar Transport." *Biological Sciences in Space* 14, no. 2 (2000): 47–57.

[17] M.J. Correll, T.P. Pyle, K.D.L. Millar, Y. Sun, J. Yao, R.E. Edelmann, and J.Z. Kiss. "Transcriptome Analyses of Arabidopsis Thaliana Seedlings Grown in Space: Implications for Gravity-Responsive Genes. *Planta* (2013): 1–15.

[18] Paul, A.-L., A. Zupanska, E. Schultz, and R. Ferl. "Organ-Specific Remodeling of the Arabidopsis Transcriptome in response to spaceflight." *BMC Plant Biology* 13, no. 1 (2013): 112.

[19] Aubry-Hivet, D., H. Nziengui, K. Rapp, O. Oliveira, I.A. Paponov, Y. Li, J. Hauslage, N. Vagt, M. Braun, F.A. Ditengou, A. Dovzhenko, and K. Palme. "Analysis of Gene Expression During Parabolic Flights Reveals Distinct Early Gravity Responses in Arabidopsis roots." *Plant Biology* 16 (2014): 129–141.

[20] Kraft, T.F.B., J.J.W.A. van Loon, and J.Z. Kiss. "Plastid Position in Arabidopsis Columella Cells is Similar in Microgravity and on a Random-Position Machine." *Planta* 211 (2000): 415–422.

[21] Perrot-Rechenmann, C. "Cellular Responses to Auxin: Division Versus Expansion." *Cold Spring Harbor Perspectives in Biology* 2, no. 5 (2010): a001446.

[22] Yu, F., D. Driss-École, J. Rembur, V. Legué, and G. Perbal. "Effect of Microgravity on the Cell Cycle in the Lentil Root." *Physiologia Plantarum* 105 (1999): 171–178.

[23] Matía, I., F. González-Camacho, R. Herranz, J.Z. Kiss, G. Gasset, J.J.W.A. van Loon, R. Marco, and F.J. Medina. "Plant Cell Proliferation and Growth are Altered by Microgravity Conditions in Spaceflight." *Journal of Plant Physiology* 167 no. 3 (2010): 184–193.

[24] Mizukami, Y. "A Matter of Size: Developmental Control of Organ Size in Plants." *Current Opinion in Plant Biology* 4, no. 6 (2001): 533–539.

[25] Paul, A.-L., C. Amalfitano, and R. Ferl. "Plant Growth Strategies are Remodeled by Spaceflight. *BMC Plant Biology* 12, no. 1 (2012): 232.

[26] Manzano, A., J. van Loon, P. Christianen, J. Gonzalez-Rubio, F.J. Medina, and R. Herranz. "Gravitational and Magnetic Field Variations Synergize to Cause Subtle Variations in the Global Transcriptional State of Arabidopsis In Vitro Callus Cultures." *BMC Genomics* 13, no. 1 (2012): 105.

[27] Martzivanou, M., M. Babbick, M. Cogoli-Greuter, and R. Hampp. "Microgravity-Related Changes in Gene Expression After Short-Term Exposure of Arabidopsis Thaliana Cell Cultures." *Protoplasma* 229, no. 2–4 (2006): 155–162.

[28] Millar, K.D.L., P. Kumar, M.J. Correll, J.L. Mullen, R.P. Hangarter, R.E. Edelmann, and J.Z. Kiss. "A Novel Phototropic Response to Red Light is Revealed in Microgravity." *New Phytologist* 186, no. 3 (2010): 648–656.

[29] Machida, Y., H. Fukaki, and T. Araki. "Plant Meristems and Organogenesis: The New Era of Plant Developmental Research." *Plant Cell Physiol* 54, no. 3 (2013): 295–301.

[30] De Micco, V., G. Aronne, G. Colla, R. Fortezza, and S. De Pascale. "Agro-Biology for Bio-Regenerative Life Support Systems in Long-Term Space

Missions: General Constraints and the Italian efforts." *Journal of Plant Interactions* 4, no. 4 (2009): 241–252.

[31] Kordyum, E.L. "Plant Cell Gravisensitivity and Adaptation to Microgravity." *Plant Biology (Stuttg)* 16, Suppl 1 (2014): 79–90.

[32] Sieberer, B.J., H. Kieft, T. Franssen-Verheijen, A.M. Emons, J.W. Vos. "Cell proliferation, cell shape, and microtubule and cellulose microfibril organization of tobacco BY-2 cells are not altered by exposure to near weightlessness in space." *Planta* 230, no. 6 (2009): 1129–1140.

[33] Skagen, E.B., and T.H. Iversen. "Simulated Weightlessness and Hyper-g Results in Opposite Effects on the Regeneration of the Cortical Microtubule Array in Protoplasts from Brassica Napus Hypocotyls." *Physiol Plant* 106, no. 3 (1999): 318–325.

[34] De Micco, V., and G. Aronne. "Biometric Anatomy of Seedlings Developed Onboard of Foton-M2 in an Automatic System Supporting Growth." *Acta Astronautica* 62 (2008): 505–513.

[35] Shevchenko, G.V., I.M. Kalinina, and E.L. Kordyum. "Interaction Between Microtubules and Microfilaments in the Elongation Zone in Arabidopsis Root Under Clinorotation." *Advances in Space Research* 39 (2007): 1171–1175.

[36] Kiss, J.Z., W.J. Katembe, and R.E. Edelmann. "Gravitropism and Development of Wild-Type and Starch-Deficient Mutants of Arabidopsis During Spaceflight." *Physiol Plant* 102, no. 4 (1998): 493–502.

[37] Levine, L.H., H.G. Levine, E.C. Stryjewski, V. Prima, and W.C. Piastuch. "Effect of Spaceflight on Isoflavonoid Accumulation in Etiolated Soybean Seedlings." *Journal of Gravitational Physiology* 8, no. 2 (2001): 21–27.

[38] Perbal, G. "The Role of Gravity in Plant Development." In *A world without gravity*, B. Fitton and B. Battrick edited by. Vol. SP-1251. European Space Agency pp. 121–136, 2001.

[39] Stutte, G.W., O. Monje, G.D. Goins, and B.C. Tripathy. Microgravity Effects on Thylakoid, Single Leaf, and Whole Canopy Photosynthesis of Dwarf Wheat. *Planta* 223, no. 1 (2005): 46–56.

[40] Kochubey, S.M., N.I. Adamchuk, E.I. Kordyum, and J.A. Guikema. "Microgravity Affects the Photosynthetic Apparatus of Brassica rapa." *Plant Biosystems* 138 (2004): 1–9.

[41] Link, B.M., S.J. Durst, W. Zhou, and B. Stankovic. "Seed-to-Seed Growth of Arabidopsis Thaliana on the International Space Station Space Life Sciences: Gravity-Related Processes in Plants." *Advances in Space Research* 31 (2003): 2237–2243.

[42] Musgrave, M.E., A. Kuang, Y. Xiao, S.C. Stout, G.E. Bingham, L.G. Briarty, M.A. Levenskikh, V.N. Sychev, and I.G. Podolski. Gravity Independence of Seed-to-Seed Cycling in Brassica rapa. *Planta* 210, no. 3 (2000): 400–406.

[43] De Micco, V., M. Scala, and G. Aronne. "Effects of Simulated Microgravity on Male Gametophyte of Prunus, Pyrus, and Brassica Species." *Protoplasma* 228, no. 1-3 (2006): 121–126.

[44] Cucinotta, F.A., M.H. Kim, L.J. Chappell, and J.L. Huff. "How Safe is Safe Enough? Radiation Risk for a Human Mission to Mars." *PLoS One* 8, no. 10 (2013): e74988.

[45] Arena, C., V. De Micco, E. Macaeva, and R. Quintens. "Space Radiation Effects on Plant and Mammalian Cells." *Acta Astronautica* 104 (2014): 419–431.

[46] De Micco, V., C. Arena, D. Pignalosa, and M. Durante. "Effects of Sparsely and Densely Ionizing Radiation on Plants." *Radiation and Environmental Biophysics* 50, no. 1 (2011): 1–19.

[47] Wei, L.J., Q. Yang, H.M. Xia, Y. Furusawa, S.H. Guan, P. Xin, and Y.Q. Sun. "Analysis of Cytogenetic Damage in Rice Seeds Induced by Energetic Heavy Ions On-Ground and After Spaceflight." *The Journal of Radiation Research* 47, no. 3-4 (2006): 273–278.

[48] Holst, R.W., and D.J. Nagel. "Radiation effects on plants." In *Plants for Environmental Studies*. Boca Raton, FL: Lewis Publishers, 1997.

[49] Wi, S.G., B.Y. Chung, J.S. Kim, J.H. Kim, M.H. Baek, J.W. Lee, and Y.S. Kim. "Effect of Gamma Irradiation on Morphological Changes and Biological Responses in Plants." *Micron* 38 (2007): 553–564.

[50] Tanaka, A., N. Shikazono, and Y. Hase. "Studies on Biological Effects of Ion Beams on Lethality, Molecular Nature of Mutation, Mutation Rate, and Spectrum of Mutation Phenotype for Mutation Breeding in Higher Plants." *The Journal of Radiation Research* 51, no. 3 (2010): 223–233.

[51] Arena, C., V. De Micco, and A. De Maio. "Growth Alteration and Leaf Biochemical Responses in Phaseolus Vulgaris Exposed to Different Doses of Ionising Radiation." *Plant Biol (Stuttg)* 16, Suppl 1 (2014): 194–202.

[52] Hammond, E.C., K. Bridgers, and F.D. Berry. "Germination, Growth Rates, and Electron Microscope Analysis of Tomato Seeds Flown on the LDEF." *Radiation Measurements* 26 (1996): 851–861.

[53] Kovács, E., G. Ball, and A. Nessinger. "The Effect of Irradiation on Sweet Cherries." *Acta Alimentaria* 24 (1995): 331–343.

[54] Arena, C., V. De Micco, G. Aronne, M.G. Pugliese, A. Virzo, and A. De Maio. "Response of Phaseolus vulgaris L. Plants to Low-LET Ionizing Radiation: Growth and Oxidative Stress." *Acta Astronautica* 91 (2013): 107–114.

[55] Zaka, R., C.M. Vandecasteele, and M.T. Misset. "Effects of Low Chronic Doses of Ionizing Radiation on Antioxidant Enzymes and G6PDH Activities in Stipa capillata (Poaceae)." *Journal of Experimental Botany* 53, no. 376 (2002): 1979–1987.

[56] De Micco, V., C. Arena, and G. Aronne. "Anatomical Alterations of *Phaseolus vulgaris* L. Mature Leaves Irradiated with X-rays." *Plant Biol (Stuttg)* 16, Suppl 1 (2014): 187–193.

[57] Esnault, M.A., F. Legue, and C. Chenal. "Ionizing Radiation: Advances in Plant Response." *Environmental and Experimental Botany* 68 (2010): 231–237.

[58] Paradiso, R., V. De Micco, R. Buonomo, G. Aronne, G. Barbieri, and S. De Pascale. "Soilless Cultivation of Soybean for Bioregenerative Life-Support Systems: A Literature Review and the Experience of the MELiSSA Project–Food Characterisation Phase I." *Plant Biol (Stuttg)* 16, Suppl 1 (2014): 69–78.

25

Human Systems Physiology

Nandu Goswami[1], Jerry Joseph Batzel[1] and Giovanna Valenti[2]

[1]Gravitational Physiology and Medicine Research Unit, Institute of Physiology, Medical University of Graz, Austria
[2]Department of Biosciences, Biotechnologies and Biopharmaceutics, University of Bari, Aldo Moro, Italy

25.1 Introduction

The main objective of this chapter is to examine bed rest as a ground-based analog for the effects of microgravity on integrative physiological systems as encountered in space flight. Many effects of microgravity are well-known and are reviewed elsewhere in this volume. However, as longer space flights are contemplated, it becomes ever more important to be able to carry out well-controlled repeatable studies that probe these effects and allow for countermeasure strategies to be developed to limit the negative effects of microgravity. The bed rest study protocol, involving subjects lying in supine position over a time interval, represents such a highly controllable experimental environment that can provide important opportunities to examine physiological function in response to reduced gravitational stress. Bed rest studies also allow for relatively easy implementation and testing of countermeasures to reduce the detrimental effects of microgravity.

25.2 Complications of Space-Based Physiological Research

While it is in some sense obvious, it is important to emphasize that in-flight experiments often suffer from a number of procedural complications that can impair the utilization and application of data collected during space flight and

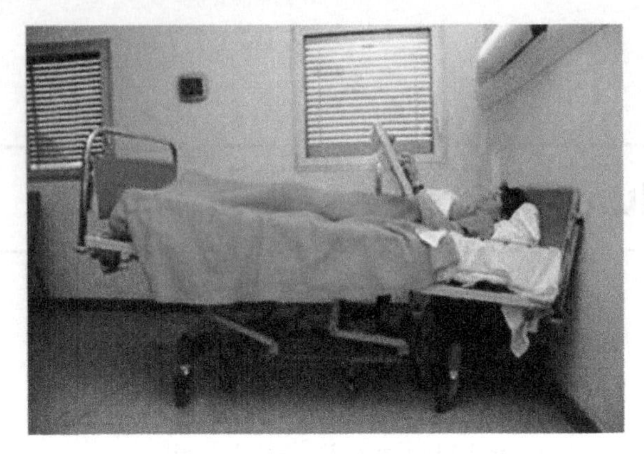

Figure 25.1 Typical configuration of the 6° head down tilt bed rest paradigm. (Image: ESA).

which can also complicate the comparison of data collected across missions [8]. These include the following:

- problems in experimental reproducibility such as limited size and uncoordinated astronaut sample populations;
- variations in quality and protocol of data measurements;
- lack of coordination in measurements across missions;
- lack of opportunity to carry out variations in experimental protocols based on new information due to restricted flight availability;
- possible usage of in-flight medication; and
- possible interference in spaceflight data from other protocols.

Therefore, there is clearly a need to have ground-based analogs of spaceflight which can be used to study spaceflight induced deconditioning and test new and novel countermeasures.

25.3 Ground-Based Analogs of Spaceflight-Induced Deconditioning: Bed Rest and Immersion

Commonly used ground-based analogs include bed rest studies and water immersion. Clearly, both bed rest studies and immersion can avoid all the above-outlined technical problems while altering various features of gravitational loading. Water immersion involves either direct contact of the body with water ("wet"), or with the body insulated from the water ("dry"). An interesting review of long-term water (dry) immersion as a model for microgravity is given in Navasiolava *et al.* (2011), which also provides some parallel comparisons to bed rest [15]. Water immersion typically allows for one hygiene period out of immersion per day and long periods of immersion must be carried out as dry immersion, which is somewhat complicated to implement regarding experimental measurements and testing of countermeasures. As a consequence, the technical problems of water immersion are more involved than for bed rest studies. In this chapter, emphasis will be placed on bed rest studies. Some description of the bed rest study protocol will be provided along with discussion of the key question as to whether the results obtained from such studies reflect those that are obtained from spaceflight. Similarities and differences between bed rest and water immersion will also be considered where relevant.

25.4 Types of Bed Rest, Durations, and Protocols

The first studies of bed rest related to space flight began soon after human space travel began. A book published in 1986 entitled "Inactivity: physiological effects" edited by Sandler and Vernikos [20] pointed to early research as taking inactivity as a model of deconditioning and as a primary paradigm for space flight effects [16]. Over time, a broader recognition of the many interacting effects of microgravity on the body viewed as a whole organism was established as was the model of simulating microgravity via head down bed rest [12, 16, 19].

Over the years, researchers working in the area of gravitational and spaceflight physiology have used subjects who were bed-rested in the supine position or at various levels of head down tilt such as 5°, 10°, or 15° (referred to here as head down bed rest, HDBR). For the details of bed rest studies that have used different angles of tilt as well as varying durations of bed rest, the reader is referred to Sandler and Vernikos [20]. To mimic the effects of microgravity, 6-degree head down supine body position is the standard implementation, which acts to equilibrate the distribution of blood and tissue volume, simulating the lack of gravitational pull to the lower body [12, 16].

Bed rest studies can be adjusted to mimic various space flight durations as well. Typical duration spans are short-term (5–7 days), medium-term (21 days), and long-term studies (60–90 days). Bed rest studies can be restricted to all male or all female studies such as a 90-day male bed rest study in 2001/2002 and the 60-day female bed rest study WISE-2005 [12].

Recently, there have been initiatives to introduce some standardization in bed rest study protocols to allow for greater cross-study exploitation of data [6, 12]. Such standardization would reduce confounding influences and allow for the comparison of different systems or different measurements under a common perturbation.

25.5 Physiological Systems Affected by Spaceflight and Bed Rest

Lack of gravity poses many challenges, which can impair the function of a number of physiological systems. On the other hand, microgravity represents a unique window for observing the response of physiological systems because it represents the suspension of those evolutionary challenges that shaped such systems. Bed rest studies allow for more delicate, complex, and even invasive measurements, which may be difficult or impossible to implement or to accurately carry out in space such as bone marrow biopsies or complex sonographies.

Direct information on microgravitational effects in space flight and information from bed rest studies reenforce the understanding of effects in both areas, and together, both sources of information can provide further novel information on physiological system function in general and in specific processes. For example, bed rest and spaceflight are both accompanied by loss of plasma volume and hemoconcentration, thus leading to a higher risk of blood clotting with potential dire consequences. We are not aware, however, of any study that examined clotting changes during spaceflight. On the other hand, there are several studies, which have examined the effects of physical inactivity during bed rest on clotting [9].

Similarly, recovery from bed rest immobilization provides an important surrogate for astronaut recovery after flight as well. Deconditioning following either weightlessness or bed rest shares many effects and symptoms characterized by orthostatic intolerance (OI) [4]. Hence, studies that reflect on the origin of OI and the efficacy of countermeasures in the bed rest context can be expected to translate into countermeasures in space. The review by Convertino [4] provides useful examples of this observation. In addition, studies in both microgravity and bed rest conditions can provide information on the physiology and clinical problems related to OI and potential countermeasures for patients suffering from OI.

A great deal of research related to physiological effects of microgravity in space and simulated microgravity during bed rest has been published, and review articles summarizing this research have appeared as well. In this section, we mention several reviews in specific areas as well as general reviews, which cumulatively set the context for taking bed rest as a valid surrogate for space microgravity. We now discuss some key physiological systems where important parallels can be made between the physiological effects of microgravity and bed rest. These areas include the autonomic nervous system, musculo-skeletal system, cardiovascular system, neuro-vestibular system, immune system, and the renal system.

Autonomic nervous system (ANS): The ANS is involved in the control of cardiovascular, thermoregulatory, respiratory, metabolic, gastrointestinal, and many other systems, and gravity influences these systems not only directly but also through feedback control mechanisms involving autonomic function which are thus impacted by a lack of gravitational stress from multiple sources. For example, sympathetic-vestibular links to blood pressure control are influenced by bedrest [7] as have effects of microgravity on baroreceptor sensitivity [1].

Musculo-skeletal system: Due to a lack of gravitational loading on the musculoskeletal system, muscle loss and bone loss occur during bed rest and spaceflight [10]. Indeed, bed rest has been recognized as a very useful and appropriate model for bone loss observed in space flight and microgravity [6, 13, 24]. The effects during bed rest on bone mineral density, bone markers, and calcium balance, and excretion are qualitatively consistent (although of lesser magnitude) to those deleterious effects seen in space flight [13]. Bed rest has also been used to test countermeasures to mitigate these effects. The review by Leblanc *et al* (2007) provides an extensive comparison and assessment of bed rest and space flight data on bone loss [13].

Cardiovascular system: The bed rest model can very effectively reflect the effects of microgravity on the cardiovascular system and in particular cardiovascular deconditioning and orthostatic intolerance (see [4]). The effect of reduction in plasma volume induced by microgravity impacts cardiac function in a variety of ways and reduces baroreflex sensitivity as well [24]. Many of these effects can also be seen in HDBR (see [16]), indicating the appropriateness of this model for studying the effects of microgravity as well as potentially linking these effects with the deconditioning in aging (see below).

Neurovestibular system: This system integrates mechanisms related to posture, eye movements, spatial orientation, and higher cognitive processes such as 3-D vision. This system also plays a role in the respiratory and cardiovascular systems, as well as many other regulatory mechanisms such as circadian regulation. Weightlessness impacts vestibular function in many ways, and HDBR may also induce related effects. The study by Dyckman *et al* (2012) indicates an impact of bed rest on the vestibulo-sympathetic response, which may influence blood pressure control and hence play a role in orthostatic intolerance [7]. This study illustrates the point that, given the complexity of interacting systems, the parallels between microgravity and head down bed rest need to be carefully considered when drawing inferences in specific scenarios.

Immune system: This complex system is certainly influenced by spaceflight and can be be affected by many factors including stress and exposure to radiation [6]. Bed rest studies related to immune system effects can differ from effects seen in space [6]. Hence, it is not clear how appropriate the bed rest model is for immune effects although it can serve as a comparison for other models [6]. The immune system response highlights again the fact that care must be taken when considering parallels between space flight microgravity and bed rest.

Renal function, volume regulation and aquaporins: Body water balance is regulated by vasopressin. Vasopressin signaling promotes water reabsorption in the renal collecting duct by triggering redistribution of the water channels Aquaporin-2 (AQP2) from intracellular vesicles into the plasma membrane [23]. AQP2 is partially excreted in the urine [22] and can represent a useful noninvasive biomarker for understanding the physiological renal response (and adaptation) to alteration in external gravity. AQP2 excretion has been evaluated as a biomarker of renal adaptation to microgravity in both water immersion and HDBR models [22, 23].

In addition to the studies and reviews on specific physiological topics referenced above, several very useful general reviews have been published. Pavy-Le Traon *et al* (2007) provide a comprehensive review of research over the last 20 years of important areas where the effects of bed rest and microgravity overlap [16]. This review provides a convincing case for viewing bed rest as a powerful research tool and useful analog for studying the effects of microgravity. The review by Vernikos and Schneider (2010) presents many parallels between physiological effects of bed rest, microgravity, and aging, providing clear insight into how these three physiological conditions can be merged and coordinated to provide a more global view of physiological function [24]. The review by Navasiolava *et al* (2011) provides a comparative look of bed rest and water immersion [15].

25.6 Is Bed Rest a Valid Analog for Microgravity-Induced Changes?

A key issue is the degree to which bed rest represents a useful surrogate for microgravity, leading to knowledge reflecting the physiological responses to microgravity in space flight and information transferable to the design of effective countermeasures against the negative effects of microgravity. HDBR (typically at 6°) is the most commonly used protocol to mimic microgravity especially for long-duration studies [5] and in relation to complex physiological effects [6]. Horizontal supine bed rest immobilization has also been used to simulate spaceflight induced deconditioning. While both horizontal supine bed rest and HDBR cause deconditioning, cephalad fluid shifts, and the onset of hemodynamic changes appear faster during HDBR (detailed in [20]).

The acceptance of HDBR as a valid surrogate for global and specific micro-gravitational influences on human physiology emerged as a consequence of carrying out many bed rest studies focusing on specific physiological areas over a wide range of systems and comparing results to direct physiological studies of astronauts and cosmonauts. For example, OI is an important problem for returning astronauts [2] and bed rest subjects face similar problems at the end of the experimental bed rest procedure. The review of the applicability of bed rest data to the problem of OI after space flight given in [4] illustrates the usefulness of the comparison between these two manifestations of OI.

While HDBR has been widely accepted as a valid surrogate protocol for microgravity in studying physiological systems, it is important to note that the information from bed rest studies needs to be carefully assessed in its application to microgravity given that bed rest is not a perfect parallel for microgravity. For example, some research suggests that certain immune system responses are affected by stresses seen in space flight rather than as a direct result of microgravity [17, 24]. Hence, specific immune system effects may not always arise in bed rest-simulated microgravity [6] (see also discussion above on immune system).

In addition, physiological responses can vary between ground-based analogs. As mentioned above, AQP2 excretion has been evaluated as a biomarker of renal adaptation to microgravity in both water immersion and HDBR models [22, 23]. These studies revealed that while AQP2 excretion is probably not a good biomarker to monitor renal fluid regulation during acute water immersion, it could instead represent a reliable and informative parameter during prolonged bed rest and possibly under chronic adaptation to microgravity in space. Specifically, it has been found that bed rest induces a biphasic response in AQP2 excretion which is in agreement with expected fluid redistribution in microgravity: in the early phase the decrease in AQP2 excretion is paralleled by an increase in hematocrit due to reduction in plasma volume, whereas the subsequent increase in AQP2 excretion is paralleled by a partial restoration of hematocrit likely due to an AQP2-mediated water reabsorption in an attempt to restore normal plasma volume. The increase in AQP2 excretion may be consequent to an increase in vasopressin (because of reduced plasma volume) as also observed in astronauts during microgravity. These results make urinary AQP2 excretion a reliable biomarker of renal water handling during prolonged bed rest and more appropriate in ground based models to mimic chronic adaptation to microgravity.

Confounding factors and interrelated effects that could differ between the microgravity of space and the simulated microgravity of bed rest (or water immersion) may be very difficult to analyze. For example, vestibular control depends on position, orientation, movement, vision, and gravitational signals, and interacting effects may be very complex as seen in a study of vestibular-sympathetic response and its relation to OI in short-term and long-term bed rest [7]. One needs to clearly establish whether, when and how effects differ between a fixed supine position mimicking microgravity compared to moving astronauts experiencing direct microgravity. Indeed, differences can arise in the physiological effects among the three situations of true microgravity, HDBR, and water immersion. For example, each of these three situations involves differing influences on tissue and vascular distribution of fluids [19]. Hence, for effective modeling of microgravity effects (or the effects of aging for that matter), one needs to consider whether any differences between true microgravity and microgravity surrogates can be ignored when drawing parallels for the specific phyiological mechanisms under study and when these differences may distort conclusions.

25.7 Bed Rest: A Testing Platform for Application of Countermeasures to Alleviate Effects of Microgravity—Induced Deconditioning

Countermeasures to microgravitational effects can be designed and incorporated in bed rest studies to test many potential approaches including such novel methods as short-arm centrifuge, [3] which in addition can incorporates ergometric exercise [14]. Such methods can be tested for preventing or reducing spaceflight deconditioning impairing cardiovascular, bone metabolism, musculoskeletal, autonomic, and other system functions. Information derived from tests of possible countermeasures for spaceflight deconditioning will be critical for microgravity exposures of long duration as will be encountered, for example, during a mission to Mars. An example of analysis related to deconditioning and aerobic exercise during bed rest can be found in Lee *et al* (2010) [14], and the implementation of centrifugation as a countermeasure implemented via bed rest is illustrated in Clément and Pavy-Le Traon (2004) [3]. Combined exercise and nutrition countermeasures have been discussed in Schneider *et al* (2009) [21].

25.8 Perspectives

In addition to the direct application to space flight, bed rest and other immobilization studies have been designed and carried out to study many other physiological conditions related to a diverse set of situations such as hospital demobilization after surgery or injury, coma, paralysis, as well as to study the effects that arise due to aging. It is important to note that there is a potential for developing a powerful synergy of information relating broader bed rest problems and the problem of microgravity. In particular, there are clear potential parallels between microgravity, aging, and immobilization [24].

Bed rest (and water immersion) are global models for the effects of microgravity, influencing a number of systems simultaneously. This is an important advantage but it is also possible to consider models of microgravity that influence individual systems or specific system levels. For example, specialized devices have been used for studying the effects of reduced gravity at the cellular level [11]. Hence, one always needs to keep in mind the research goals when considering appropriate surrogates for physiological effects of microgravity and space flight.

References

[1] Natalia M. Arzeno, Michael B. Stenger, Stuart M. C. Lee, Robert Ploutz-Snyder and Steven H. Platts. Sex differences in blood pressure control during 6° head-down tilt bed rest. American Journal of Physiology Heart Circulation Physiology, 15, 304, 8: H1114–1123, 2013.

[2] Andrew P. Blaber, Nandu Goswami, Roberta L. Bondar and Mohammed S. Kassam. Impairment of cerebral blood flow regulation in astronauts with orthostatic intolerance after flight. Stroke, 42, 7:1844–1850, 2011.

[3] Gilles Clément and Anne Pavy-Le-Traon. Centrifugation as a countermeasure during actual and simulated microgravity: a review. European Journal of Applied Physiology, 92, 3: 235–248, 2004.

[4] Victor A. Convertino. Insight into mechanisms of reduced orthostatic performance after exposure to microgravity: comparison of ground-based and space flight data. Journal of Gravitational Physiology, 5, 1: 85–88, 1998.

[5] Victor A. Convertino and Caroline A. Rickards. Human models of space physiology. Chapter 48 In: Conn, P. Michael (ed,) Source book of models for biomedical research, Humana Press- Springer, Trenton, pp 457–464, 2008.

[6] Brian E. Crucian, Raymond P. Stowe, Satish K. Mehta, Deborah L. Yetman, Melanie J. Leal, Heather D. Quiriarte, Duane L. Pierson and Clarence F. Sams. Immune status, latent viral reactivation, and stress during long-duration head-down bed rest. Aviation Space Environmental Medicine, 80(5 Suppl): A37–44, 2009.

[7] Damian J. Dyckman, Charity L. Sauder and Chester A. Ray. Effects of short-term and prolonged bed rest on the vastibulosympathetic reflex. American Journal of Physiology Heart Circulation Physiology 302, 1: H368–374, 2012.

[8] Nandu Goswami, Jerry J. Batzel, Gilles Clement, Peter T. Stein, Alan R. Hargens, Keith M. Sharp, Andrew P. Blaber, Peter G. Roma and Helmut G. Hinghofer-Szalkay. Maximizing information from space data resources: a case for expanding integration across research disciplines. European Journal of Applied Physiology, 113: 1645–1654, 2013.

[9] Gerhard Cvirn, James Elvis Waha, Gerhard Ledinski, Axel Schlagenhauf, Bettina Leschnik, Martin Koestenberger, Erwin Tafeit, Helmut Hinghofer-Szalkay and Nandu Goswami. Bed rest does not induce hypercoagulability. European Journal of Clinical Invesigation. 2014 Nov 21. doi: 10.1111/eci.12383.

[10] Alan R. Hargens, Roshmi Bhattacharya and Susanne M. Schneider. Space physiology VI: exercise, artificial gravity, and countermeasure development for prolonged space flight. European Journal of Applied Physiology, 113: 2183–2192, 2013.

[11] Lifang Hu, Runzhi Li, Peihong Su, Yasir Arfat, Ge Zhang, Peng Shang and Airong Qian. Response and adaptation of bone cells to simulated microgravity. Acta Astronautica, 104(1), 396–408, 2014.

[12] Peter D. Jost. Simulating human space physiology with bed rest. Hippokratia suppl 1: 37–40, 2008.

[13] Adrian D. LeBlanc, Elisabeth R. Spector, Harlan J. Evans and Jean D. Sibonga. Skeletal responses to space flight and the bed rest analog: a review. Journal of Musculoskeletal Neuronal Interactions, 7,1: 33–47, 2007.

[14] Stuart M. C. Lee, Alan D. Moore, Meghan E. Everett, Michael B. Stenger and Steven H. Platts. Aerobic exercise deconditioning and countermeasures during bed rest. Aviation Space Environmental Medicine, 81,1, 52–63, 2010.

[15] Nastassia M. Navasiolava, Marc-Antoine Custaud, Elena S. Tomilovskaya, Irina M. Larina, Tadaaki Mano, Guillemette Gauquelin-Koch, Claude Gharib and Inesa B. Kozlovskaya. Long-term dry immersion:review and prospects. European Journal of Applied Physiology, 111, 7: 1235–1260, 2011.

[16] Anne Pavy-Le Traon, Martina Heer, Marco Narici, Joern Rittweger and Joan Vernikos. From space to earth: advances in human physiology from 20 years of bed rest studies (1986–2006). European Journal of Applied Physiology, 101: 143–194, 2007.

[17] Duane L. Pierson, Satish K. Mehta and Raymond P. Stowe. Reactivation of latent herpes viruses in astronauts. In: Psychoneuroimmunology (4th ed.), edited by Ader R., Felten D. L. and Cohen N. Philadelphia, PA: Elsevier, Inc., p. 851–868, 2007.

[18] Valerie V. Polyakov, Natasha G. Lacota and Alexander Gundel. Human thermohomeostasis onboard "Mir" and in simulated microgravity studies. Acta Astronauta, 49: 137–43, 2001.

[19] Jacques Regnard, Martina Heer, Christian Drummer and Peter Norsk. Validity of microgravity simulation models on earth. American Journal of Kidney Disease, 38, 3, 668–674, 2001.

[20] Harold Sandler and Joan Vernikos (eds.). Inactivity: physiological effects. Academic Press, New York, 1986.

[21] Suzanne M. Schneider, Stuart M. C. Lee, Brandon R. Macias, Donald E. Watenpaugh and Alan R. Hargens. WISE-2005: exercise and nutrition countermeasures for upright VO2pk during bed rest. Medicine Science Sports and Exercise, 41: 2165–2176, 2009.

[22] Grazia Tamma, Annarita Di Mise, Marianna Ranieri, Maria Svelto, Rado Pisot, Giancarlo Bilancio, Pierpaolo Cavallo, Natale G. De Santo, Massimo Cirillo and Giovanna Valenti. A decrease in aquaporin 2 excretion is associated with bed rest induced high calciuria, Journal of Translational Medicine, 12, 133–134, 2014.

[23] Giovanna Valenti, Walter Fraszl, F. Addabbo, Grazia Tamma, Procino G., E. Satta, Massimo Cirillo, Natale G. De Santo, Christian Drummer, L. Bellini L. *et al.*, Water immersion is associated with an increase in aquaporin-2 excretion in healthy volunteers. Biochim Biophys Acta, 1758, 8, 1111–1116, 2006.

[24] Joan Vernikos and Suzanne V. Schneider. Space, gravity and the physiology of aging: parallel or convergent disciplines? A mini-review. Gerontology, 56, 2:157–166, 2010.

26

Behavior, Confinement, and Isolation

Carole Tafforin

Ethospace, Toulouse, France

With long-term space missions, behavioral scientists agree that the personal and interpersonal adaptive processes became of prime importance for crew performance in isolation and confinement. Anecdotal reports from crews of the Russian orbital station (Mir) and the International Space Station (ISS) indicated such psychosocial issues. The crewmembers thus have to adapt to a wide range of environmental factors: reduced and closed space, life-support restriction, group density, delayed communication, far from civilization, lack of variety of food, lack of natural light and surrounding, lack of privacy, and monotony of daily life. These include factors related to space crew heterogeneity: interpersonality, crew demographics, value, culture, and language background. Effective procedures for selection, training, and psychological support will help to determine the outcome of next space missions and space explorations. Psychosocial adaptation to a Mars mission is a new challenge [1] and involves temporal factor in a novel way. Time has definitely an impact on behavior. For simulating those factors, generating on Earth an extra-terrestrial environment would help in getting better knowledge of human factors with a focus on the behavior of confined and isolated crews.

Researches specifically relevant to *confined crews* concern undersea habitats including submarines campaigns and closed-tanks experiments in multi-chamber facilities (i.e., the space simulators).

The current results in Capsule habitats state a positive psychology in terms of salutogenic adaptation [2] although inter-individual conflicts, changes of mood, high levels of tension/anxiety, depression and mental disturbances due to deprivations, dangers, and stresses were detected. They have an impact on coping style, efficiency at work, motivation to the mission goal, cognitive functioning, crew cohesion, and crewmember's well-being as a result.

During NEEMO[1] missions for instance, the individuals that lived inside the divers' undersea habitat with strict life-support and worked outside for testing equipment in buoyancy, experienced a powerful training that have an enduring value, a benefic effect, and an optimized success for their allotted space missions. High crewmember autonomy would become the norm for future exploratory missions [3].

During submarine missions, the crew is confronted to additional environmental properties: crowding, schedule per quarters, absence of day/night cues, no communication with Earth, no information on underway position, and prolonged dangerous operations. The rigorous and very long crew training that occurs before military deployments and the presence of strong hierarchical rules lead to behavioral stereotypes and over-learned technical responses [4] that do not permit uncertain outcomes in the adaptive process.

Closed tank experiments were justly designed with the aim of studying the crewmembers' behavioral strategies of adaptation. Ethological investigations made during ISEMSI[2], EXEMSI[3], and HUBES[4] experiments described changes in social behavior over time and according to the environmental situation [5]. Within large or open areas, interindividual distances were constant. In reduced habitats, the frequency of personal distances, according to Hall's classification, decreased and the frequency of public distances increased with high levels of social distance and body mobility from the initial period to the final period of confinement. Over the campaigns, living and working together in closed tanks were more and more stressful and the long-term adaptive process was not still achieved after 135 days. In the SFINCSS[5] experiment, comparing one group confined for 240 days and two groups for 110 days, differences in culture and attitudes toward gender were factors identified as having a major impact on the intergroup relationship [6]. Other results indicated coping behaviors. Environmental stresses were identified altering well-being and human performance inside the multi-chamber facilities.

The Mars-500 experiment offered an exceptional paradigm that promoted further advances on crew behavior not only in confinement but also in extended time period and in high autonomy. Summarized results from multidisciplinary

---

[1]NASA Extreme Environment Mission Operations.
[2]Isolation Study for European Manned Space Infrastructure.
[3]EXperimental campaign for European Manned Space Infrastructure.
[4]HUman Behavior in Extended Space flight.
[5]Simulation of Flight International Crew on Space Station.

approaches [7] showed a progressive reduction in physical activities during the course of the simulation, disturbances in sleep quality and quantity, changes in patterns of language within the high autonomy phase and during Mars landing period, increased perceived homogeneity in personal values, consistency in the most salient personal goals, gain of positive strengths from demanding situation, and increased loneliness particularly at the end of 520-day confinement. An ethological monitoring of the crew globally pointed out time effects, cultural preferences and individual differences in the crew's actions, interactions, expressions and communications during Mars-500 experiment. A personal account by the crewmembers who experienced this simulated interplanetary flight reported, for some of them, difficult periods when few contacts were arriving from outside, and for all of them, the lack of connection with nature, the lack of fresh and variety of food, and the importance of communication channels. Psychological crew support program, implemented as countermeasure in the experiment, provided efficient communication sessions.

Such studies have demonstrated the interest to extend ground simulations of the psychosocial environment encountered during very long-duration stays in isolation and confinement. This conducted to a focus on the isolation factor.

Researches specifically relevant to *isolated crews* concern polar regions stays in Antarctica and in Arctic and dessertic lands (i.e., the analogue environments).

The most salient data collected from numerous winterovers in South Pole stations led to propose four characteristics regarding psychosocial adaptation to isolation in extreme settings: adaptation is *situational*, adaptation is *social*, adaptation is *cyclic*, and adaptation is *salutogenic* according to Palinkas cited in [8]. However, a significant increase in depressed mood was emphasized in men and women who spent a year at Mac Murdo Station and Amundson-Scott station in Antarctica [9], for example. All-female crews are rare and mixed-gender expeditions were developed. This provides evidence of the heterogeneous groundwork of a crew to cope, regulate, and adapt for a better equilibrium in isolation conditions. Considering multi-cultural crews, cross-cultural comparisons have provided some findings that suggest a characteristic personality trait profile in the Antarctic expeditioner [10] that may be considered in the future to select space explorer.

Studies performed at the FMAR[6] in Devon Island, indicated differences between individual coping styles across time. Stress decreased for females

[6]Flashline Mars Arctic Research Station.

while it increased for males who demonstrated higher levels of excitement, tiredness, and loneliness [11]. The results concluded that simulations of prolonged real isolation and hostile natural environment appear to provoke true demands for adaptation that actually approaches interplanetary missions. Depending on weather conditions, the crew stays for short-term (mimicking Moon landing) or for long-term missions (mimicking Mars landing). In both cases, the adaptation is *situational.*

As test beds for field operations, studies at the MDRS[7] in Utah desert investigated crew selection protocols, tested key habitat design features, replicated space food packages, compared mission crews with backup crews, analyzed the high workload on crew vigilance, and examined communications in multi-lingual crews, among other investigations. Preliminary works on language skills revealed that verbal and nonverbal expressions were influenced by the cultural background such as native language, by their respective roles within the crew and the crewmembers' spirit [12]. They contributed to show that the adaptation is *social.*

The most recently built station in Polar Regions was Concordia at Dome C. This South Pole base serves as research laboratories with a particular interest on space life sciences. The crews who stay there are considered as winterers but also as interplanetary crewmembers because of similarities with the personnel composition, architectural structure, and temporal scale. With an emphasis on the long-duration factor, ethological observations made weekly during a collective activity at meal times, allowed to describe and quantified certain profiles of social behavior according to the mission day. Interesting data were on the collective attendance and collective time. The results showed for instance, periodic changes in the number of winterers attending the meals, over three periods of 13 weeks each (Figure 26.1) and cyclic variations in the time spent altogether at meals, every 7 weeks (Figure 26.2). A Mars mission scenario on different crew organizations over time could be applied in real mission. These findings further demonstrated that the adaptation is *cyclic.*

With the opportunity of Tara expedition in Arctic, the environmental properties became more and more stressful because of the combined isolation and confinement factors in synergy with the associated periods of winterover totalizing 507 days in duration like a Mars mission. Preliminary results [13] showed that the irregularity of collective time and the variations of inter-individual positions were behavioral indicators that would prevent the

[7]Mars Desert Research Station.

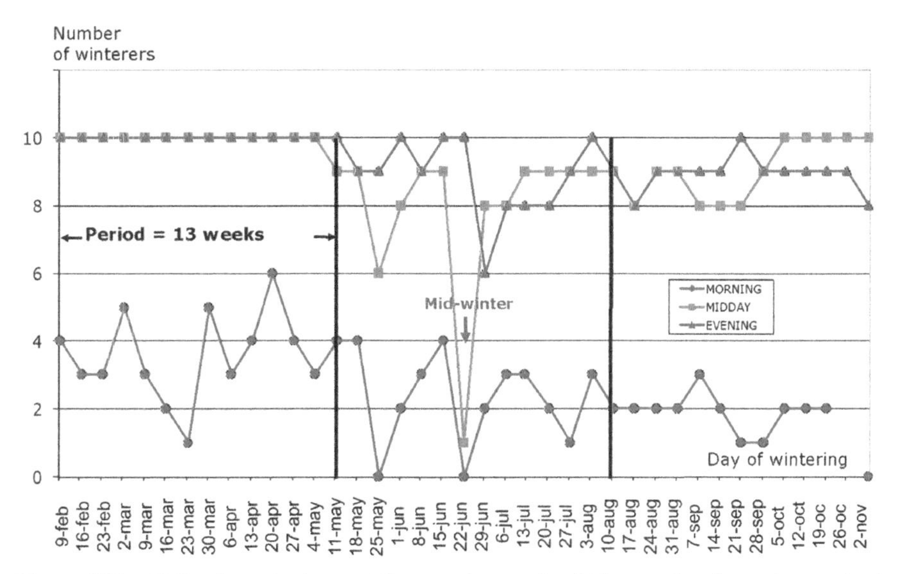

Figure 26.1 Collective attendance at the morning meal, midday meal and evening meal, at Concordia station in Antarctic, according to the days (mission DC2, 2006).

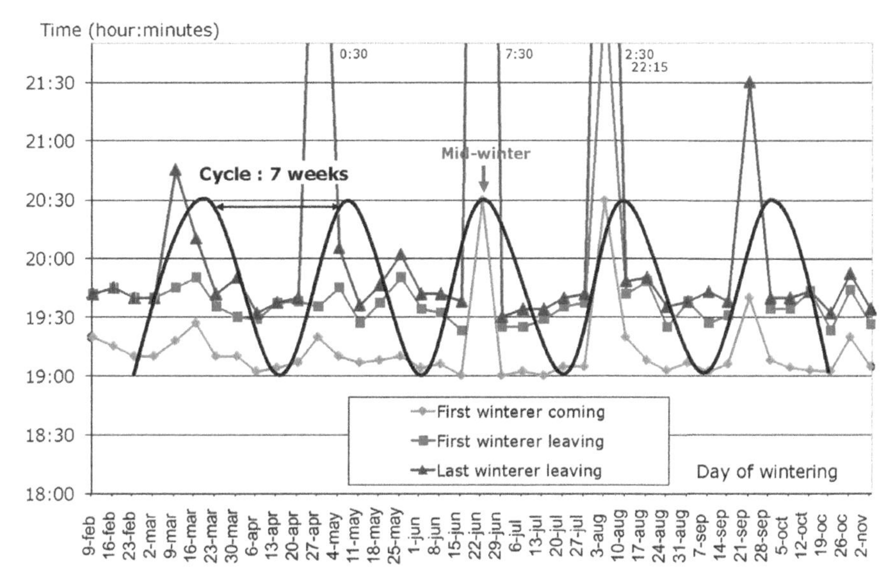

Figure 26.2 Collective time at the evening meal, at Concordia station in Antarctic, according to the days (mission DC2, 2006).

monotony of daily life. A complete study is in process. It should support the statement that the adaptation is *salutogenic*.

These examples highlight the key issues relevant to researches on isolated and confined crews' behavior in order to provide the space crews with the best quality of life and success of missions. A synthesis on psychology of space exploration exhaustively reviewed contributions in the area [8].

Future researches might include further studies on cognitive issues in terms of new crewmember's representation of the crew and in terms of new crewmember's representation of the outsight.

Acknowledgement

The author would like to thank the French Space Agency for its financial support.

References

[1] Fiedler, Edna R., and Albert A. Harrison. "Psychosocial Adaptation to a Mars Mission." *Journal of Cosmology* 12 (2010): 3694–3710.

[2] Suedfeld, Peter and G. Daniel Steel. "The Environmental Psychology of Capsule Habitats." *Annual Review of Psychology*, 51 (2000): 227–253.

[3] Eid, Jarle, Bjorn H. Johnsen, Evelyn-Rose Saus, and Jan Risberg. "Stress and Coping in a Week-Long Disabled Submarine Exercise." *Aviation Space and Environmental Medicine* 75, no. 7 (2004): 616–621.

[4] Kanas, Nick, Stephanie Saylor, Matthew Harris, et al. "High Versus Low Crewmember Autonomy in Space Simulation Environments." *Acta Astronautica* 67 (2010): 731–738.

[5] Tafforin, Carole. "Ethological Indicators of Isolated and Confined Teams in the Perspective of Missions to Mars." *Aviation Space and Environmental Medicine* 76 (2005): 1083–1087.

[6] Sandal, Gro. "Culture and Tension During an International Space Station Simulation: Results from SFINCSS'99." *Aviation Space and Environmental Medicine* 75, no. 7 (2004): C44–C51.

[7] Gushin, Vadim, "Sessions on Psychological and Psycho-Physiological Issues in Extended Space Flight and Isolation." *Mars-500 International Symposium*. Moscow, Russia, April 2012.

[8] Vakoch, Douglas A. "Psychology of Space Exploration. Contemporary Research in Historical Perspective." The NASA History Series, Washington DC, NASA SP-2011-4411, 2011.

[9] Palinkas, Lawrence A., Jeffrey C. Johnson, and James S. Boster. "Social Support and Depressed Mood in Isolated and Confined Environments." *Acta Astronautica* 54 (2004): 639–647.

[10] Musson, David M., Gro M. Sandal, Michael L. Harper, and Robert L. Helmreich. "Personality Testing in Antarctic Expeditioners; Cross Cultural Camparisons and Evidence for Generalizability." *Proceeding of 53rd International Astronautical Congress*. The World Space Congress, Houston TX, United States. # IAC-02-G.4.08, October 2002.

[11] Bishop, Sheryl L., Ryan Kobrick, Melissa Battler, and Kim Binsted. "FMARS 2007: Stress and Coping in an Arctic Mars Simulation." *Acta Astronautica* 66 (2009): 1353–1367.

[12] Mikolajczak, Marie, Carole Tafforin, and Bernard H. Foing. "Verbal and Non Verbal Communication." *EuroMoonMars campaign of crew 92. IAF International Global Lunar Conference*, Beijing, China, June 2010.

[13] Tafforin, Carole. "The Ethological Approach as a New Way of Investigating Behavioural Health in the Arctic." *International Journal of Circumpolar Health* 70, no. 2 (2011): 109–112.

Conclusions

Daniel A. Beysens[1] and Jack J. W. A. van Loon[2]

[1]CEA-Grenoble and ESPCI-Paris-Tech, Paris, France
[2]VUmc, VU-University, Amsterdam, The Netherlands

All the means that have been considered in this book exhibit advantages and inconvenience for recreating the space conditions on Earth. In order to help the reader to discern what the best means are for a given investigation, we have listed and summarized in Table these means, their advantages, and drawbacks for particular studies.

Despite the inconveniences and artifacts present for the various ground-based systems, a wealth of knowledge has been emerged from such facilities. These studies either were published as self-standing experiments or were reported as part of the preparatory process of an in-flight experiment. In any case, it is recommended to verify, in due time, any ground-based microgravity or radiation simulation observation by in-flight studies. Often, also ground-based hypergravity studies add to our overall understanding of the effect of weight onto systems.

Table Main advantages and unconvenients of the current means used to recreate space conditions on Earth

Mean	*Advantages*	*Inconveniences*
Drop tower	High micro-g quality	Time (∼9 s)
Parabolic flight (plane)	Easy access Human subjects Tunable g	Time (∼20 s) g-jitter (10^{-2} g)
Parabolic flight (sounding rocket)	High micro-g quality Tunable g	Time (2–14 min.)
Magnetic levitation	Long Term Easy access	g-compensation depends on sample size Possible biological effects
Plateau method	Easy	Liquid mixtures only
Centrifuge	Easy access Human subjects	g-gradients inertial shears
Tail suspension	Easy access	Animal stress Only rodents
Bed rest	Easy access	Subset of organ systems
Clinostat/ RPM	Easy	Limited volumes Fluid shear / movement stress
Vibrations	Easy	Temperature gradient
Radiation	Easy	Limited particle spectrum/ energies
Analogs	Easy Human subjects	Only subset of actual flight stress
Atmosphere chambers	Easy	Size/radiation spectrum

Index

Editor's Biographies

Dr. Daniel A. Beysens is Director of Researches at Ecole Supérieure de Physique et Chimie Industrielle - Paris Tech where he leads a research team and scientific consultant at Commissariat à l'Energie Atomique at Grenoble (France) where he was Head of Institute from 1995 to 1999. He was between 2003 and 2007 President of the European Low Gravity Research Association www.elgra.org and is since 1999 the President - founder of the OPUR International Organization for Dew Utilization www.opur.fr. His area of expertise is phase transition: in space, improve the management of fluids, with emphasis on near and supercritical conditions and on Earth, obtain water from the air by passive radiative condensation. He is the author of more than 400 publications and 9 books in the above fields of research. The Physical Society (SFP) gave him the Ancel Prize of Condensed Matter Physics in 1985, he was decorated in 1995 as a knight of Palmes Académiques for outstanding data in education and research, he shared in 2000 the Grand Prix of the French Academy of science for the discovery of a new thermalization process in space, received in 2007 the Prize of Innovative Technologies for the Environment, bestowed by ADEME. In 2008 he was given the Emergence award, bestowed by the Ministry of Research, in 2012 he obtained the Loyalty award by the SDEWES Organization and in 2013 the ELGRA Medal from the European Low Gravity Association for outstanding contributions to the field of microgravity research, in particular in the understanding of supercritical fluids.

Jack van Loon, VU-University Medical Center (VUmc) and Academic Center for Dentistry (ACTA), Amsterdam, The Netherlands, has experience in space and gravity related research for more than a quarter century. In this time he was Co-I and PI of several space experiments in Shuttle, Bion, Soyuz and ISS mainly on life sciences/bone experiments, but he also collaborated with colleagues from the plant kingdom or the physical sciences arena. He was experiment coordinator for all life sciences and education experiments for the Dutch Soyuz mission DELTA with astronaut André Kuipers in 2004. He also worked for some years in industry (Bradford Engineering) responsible for the science implementation in microgravity payloads, e.g. Microgravity Science Glovebox (MSG), Life Sciences Glovebox (LSG) and the ESA EMCS experiment container. Here he also initiated the BioPack facility which flew on STS-107. Besides flight instruments he also developed instruments and payloads for ground-based research. He manages the very first large and bench-top Random Positioning Machines (RPM). He is chairman of the Dutch Microgravity Platform and was a two terms President of the European Low Gravity Research Association (ELGRA). He was involved in various Dutch and ESA educational projects. He initiated and set the requirements of the ESA Large Diameter Centrifuge (LDC) and coordinates the activities regarding the Human Hypergravity Habitat, H3, a centrifuge where humans can be exposed to mild levels of hypergravity. He authored or co-authored over 95 peer reviewed papers and was a co-editor on several books regarding space and gravity related research.

Lightning Source UK Ltd.
Milton Keynes UK
UKOW06n0356100615

253201UK00001B/9/P